1378

Plant Polymeric Carbohydrates

Plant Polymeric Carbohydrates

Edited by

Friedrich Meuser
Technical University of Berlin, Institute of Food and Fermentation Technology, Berlin, Germany

David J. Manners
Heriot-Watt University, Edinburgh, UK

Wilfried Seibel
Federal Research Centre for Cereal, Potato, and Lipid Research, Detmold and Münster, Germany

ROYAL
SOCIETY OF
CHEMISTRY

The Proceedings of an International Symposium on Plant Polymeric Carbohydrates
held in Berlin, Germany, July 1–3, 1992.

The design on the cover of this book is adapted from an idea by
Susanne Meuser.

Special Publication No. 134

ISBN 0-85186-645-X

A catalogue record for this book is available from the British Library

Published by The Royal Society of Chemistry,
Thomas Graham House, Science Park, Cambridge
CB4 4WF

Printed by Bookcraft (Bath) Ltd.

Editors' Preface

The International Symposium on Plant Polymeric Carbohydrates which was held as a satellite symposium of the International Carbohydrate Meeting for the seventh time has by now almost become a symposium in its own right inasmuch as it has succeeded in bringing together an ever increasing number of interested experts to exchange knowledge. This has been achieved by placing the emphasis on specific aspects of carbohydrate research in the selection and organization of the items on the programme. One particular attraction of the symposium is the emphasis on those topics in which equal weight is given to the lastest findings in both fundamental and applied research. By the careful selection of interrelated topics it was possible to cover a wide range of new scientific knowledge.

The aim of the symposium was to present the latest research in subbranches of the biosynthesis and structure of polymeric carbohydrates, their rheological properties, both as pure substances and in complex bonds with other natural materials, their nutritional importance with respect to their physicochemical and nutritive properties, their industrial applications in food and non-food, including those of their chemical and enzymic conversion products. In order to demonstrate the wealth of new information, the lectures were organized in such a way that papers of a general nature were combined with papers dealing with the latest scientific findings in specific areas.

The papers have been reprinted in this volume of the Proceedings which has been divided into five chapters according to the main subjects. Although these possess a certain degree of independence within the overall context of the book, the links between them are clearly recognizable, in keeping with the intentions of the organizers of the symposium which were to enable as broad a view as possible of the latest state of the art to be imparted. The second volume which will be published by the Publication Division of the Technical University of Berlin will contain the poster contributions which were given at the symposium. Allocated to the various groups of interrelated subjects, these formed an integral part of the symposium and resulted in its becoming a remarkable scientific forum which was distinguished by a particularly intense exchange of thoughts and ideas on scientific progress in this field.

We are pleased that it was possible to hold the symposium in Berlin as it enabled the over 200 participants from 20 countries to gain an impression of the ongoing historical programme. As expressed in forewords to the symposium by the Senator of Science and Research of the Senate of Berlin and the President of the Technical University of Berlin, this historical process requires a special effort to be made, also by the German scientific community. The organizers of the symposium and their staff did indeed work very hard, putting in countless hours of painstaking work. This was made possible by the generous financial assistance from both public and private funds, by the support given by their respective institutes and the advice from the Advisory Committee and the Organizing Committee. Our thanks to these bodies will be expressed elsewhere.

At this point we wish to express our thanks to all the authors for their valuable contributions and for the effort they made in the preparation of the proofs of the manuscripts without which this piece of work would not have materialized. They have thus made our work considerably easier. Nonetheless, there was still such a great deal of editorial work to be done that the exceptionally large amount of word-processing involved in putting the manuscripts into a uniform style as required by the publisher's guidelines was clearly too much for one person to deal with. We therefore sincerely appreciate the invaluable contribution of Mrs Gisela Weber of the Institute of Food and Fermentation Technology of the Technical University of Berlin, who acted as secretary to the organizers, in the development of this work. We ourselves hope that this book will prove of service to the International Scientific Community.

Friedrich Meuser
David J. Manners
Wilfried Seibel

Contents

Contributors   ix

Biosynthesis and Chemical Structure

1   Structural Studies of Plant Cell-Wall Polysaccharides Using Enzymes
    A.G.J. Voragen, H.A. Schols and H. Gruppen   3
2   A New Method of Enzymic Analysis of Amylopectin Structure
    S. Hizukuri and J. Abe   16
3   Structure-Function Relationships of ß-Glucan Hydrolases in Barley
    G.B. Fincher   26
4   Starch Synthesis in Transgenic Plants
    L. Willmitzer, J. Koßmann, B. Müller-Röber and U. Sonnewald   33

Rheology

5   Constitutive Models for Dilute and Concentrated Food
    Biopolymer Systems
    J.L. Kokini   43
6   Viscoelastic Properties of Mixed Polysaccharides Systems
    J.L. Doublier, C. Castelain and J. Lefebvre   76
7   Determination of the Density of Starches and Cereal Products as a Function
    of Temperature and Pressure
    Ch. Millauer, G. Rosa and R. Schär   86
8   Factors Affecting the Wall Slip Behaviour of Model Wheat Flour Doughs
    in Slit Die Rheometry
    J. A. Menjivar, B. van Lengerich, C.N. Chang and D. Thorniley   104

Nutrition

9   Nutritional Importance and Classification of Food Carbohydrates
    N.-G. Asp   121
10 The Glycaemic Response after Starchy Food Consumption as Affected
    by Choice of Raw Material and Processing
    I. Björck   127

11   Resistant Starch: Measurement in Foods and Physiological Role in Man
     H.N. Englyst, S.M. Kingman and J.H. Cummings   137
12   Soluble Dietary Fibre - a Useful Concept?
     I.T. Johnson   147

Industrial Uses

13   New Industrial Uses of Starch
     H. Koch, H. Röper and R. Höpcke   157
14   Properties of Small Starch Granules and their Application in Paper Coatings
     E.C. Wilhelm   180
15   Technical Applications of Galactomannans
     F. Bayerlein   191
16   Edible Fiber from Barley and Oats by Wet-Milling
     A. Lehmussaari and M.G. Lindley   203

Chemical and Enzymic Conversion

17   Solution Properties of Plant Polysaccharides as a Function of their
     Chemical Structure
     W. Burchard   215
18   Structure and Function of Barley Malt α-Amylase
     M. Søgaard, A. Kadziola, J. Abe, R. Haser and B. Svensson   233
19   Use of Extrusion Processes for Enzymic and Chemical Modifications
     of Starch
     G. Della Valle, P. Colonna, J. Tayeb and B. Vergnes   240
20   Study of Cellulolytic Hydrolysis of Furfural Process Wastes
     K. Réczey, E. László and J. Holló   252
21   Non-Fickian Diffusion with Chemical Reaction of a Penetrant with
     a Glassy Polymer: The Gas-Solid Hydroxy-Ethylation of Potato Starch
     N.J.M. Kuipers, E.J. Stamhuis and A.A.C.M. Beenackers   262
22   Characterisation of Maltodextrin Gelling by Low-Resolution NMR
     F. Schierbaum, S. Radosta, W. Vorwerg, V.P. Yuriev,
     B.B. Braudo and M.L. German   278
     Subject Index   291

Contributors

J. Abe, Kagoshima University, Faculty of Agriculture,
Department of Biochemical Science and Technology, 21-24 Korimoto 1, Kagoshima
890, Japan

N.-G. Asp, Chemical Center, University of Lund, Department of Applied Nutrition
and Food Chemistry, P.O. Box 124, 221 00 Lund, Sweden

F. J. Bayerlein, Diamalt GmbH, Special Projects, Georg-Reismüller-Str. 32,
W-8000 München 50, Germany

A.A.C.M. Beenackers, University of Groningen, Department of Chemical
Engineering, Nyenborgh 4, 9747 AG Groningen, The Netherlands

I. Björck, Chemical Center, University of Lund, Department of Applied Nutrition and
Food Chemistry,P.O. Box 124, 221 00 Lund, Sweden

B.B. Braudo, INEOS, 125080 Moscow, Russia

W. Burchard, Institute of Macromolecular Chemistry, University of Freiburg,
Stefan-Meier-Str. 31, W-7800 Freiburg, Germany

C. Castelain, INRA, Centre de Recherche Agro-Alimentaires, B.P. 527
44026 Nantes, France

C.N. Chang, Fairlawn Development Center, Extrusion Technology, Nabisco Brands
Inc., 21-11 Route 208, Fair Lawn, NJ 07410, USA

P. Colonna, INRA, Centre de Recherche Agro-Alimentaires, B.P. 527,
44026 Nantes, France

J. H. Cummings, Medical Research Council, Dunn Clinical Nutrition Centre,
100 Tennis Court Road, Cambridge CB2 1QL, UK

G. Della Valle, INRA, Centre de Recherche Agro-Alimentaires, B.P. 527,
44026 Nantes, France

J.-L. Doublier, INRA, Centre de Recherche Agro-Alimentaires, B.P. 527
44026 Nantes, France

H.N. Englyst, Medical Research Council, Dunn Clinical Nutrition Centre,
100 Tennis Court Road, Cambridge CB2 1QL, UK

G.B. Fincher, La Trobe University, Department of Biochemistry,
Victoria, Bundoora 3083, Australia

M.L. German, INEOS, 125080 Moscow, Russia

H. Gruppen, Wageningen Agricultural University, Department of Food Science,
Bomenweg 2, 6703 HD Wageningen, The Netherlands

R. Haser, LLCMB-CNRS, URA 1296, 13326 Marseille, France

S. Hizukuri, Kagoshima University, Faculty of Agriculture,
Department of Biochemical Science and Technology, 21-24 Korimoto 1, Kagoshima
890, Japan

R. Höpcke, Cerestar Deutschland GmbH, Postfach 9040, W-4150 Krefeld 12,
Germany

J. Holló, Central Research Institute for Chemistry, Hungarian Academie of Sciences,
P.O. Box 17, 1525 Budapest, Hungary

I. T. Johnson, AFRC Institute of Food Research, Norwich Laboratory,
Norwich Research Park, Colney Lane, Norwich NR4 7UA, UK

A. Kadziola, LLCMB-CNRS, URA 1296, 13326 Marseille, France

S. M. Kingman, Medical Research Council Dunn, Clinical Nutrition Centre,
100 Tennis Court Road, Cambridge CB2 1QL, UK

H. Koch, Cerestar, Research & Development, Havenstraat 84, 1800 Vilvoorde,
Belgium

J. L. Kokini, Rutgers - The State University of New Jersey,
Department of Food Science, Cook College, P.O. Box 231, New Brunswick,
New Jersey, 08903-0231, USA

J. Koßmann, Institut für Genbiologische Forschung Berlin GmbH,
Ihnestrasse 63, W-1000 Berlin 33, Germany

N.J.M. Kuipers, University of Groningen, Department of Chemical Engineering,
Nyenborgh 4, 9747 AG Groningen, The Netherlands

E. László, Technical University of Budapest, Dept. Agric. Chem. Techn.,
Gellért tér 4, 1521 Budapest, Hungary

J. Lefebvre, INRA, Centre de Recherche Agro-Alimentaires, B.P. 527
44026 Nantes, France

A. Lehmussaari, Alko Ltd., Food Division, 05200 Rajamäki, Finland

B. van Lengerich, Bühler AG, Bahnhofstrasse, 9240 Uzwil, Switzerland

J. A. Menjivar, Fairlawn Development Center, Extrusion Technology, Nabisco
Brands Inc., 21-11 Route 208, Fair Lawn, NJ 07410, USA

Ch. Millauer, Bühler AG (NM 1), Bahnhofstrasse, 9240 Uzwil, Switzerland

B. Müller-Röber, Institut für Genbiologische Forschung Berlin GmbH,
Ihnestrasse 63, W-1000 Berlin 33, Germany

S. Radosta, Fraunhofer-Institute for Applied Polymeric Research Teltow,
O-1505 Bergholz-Rehbrücke, Germany

K. Réczey, Technical University of Budapest, Dept. Agric. Chem. Techn.,
Gellért tér 4, 1521 Budapest, Hungary

H. Röper, Cerestar, Research & Development, Havenstraat 84, 1800 Vilvoorde,
Belgium

G. Rosa, Bühler AG (NM 1), Bahnhofstrasse, 9240 Uzwil, Switzerland

M. Søgaard, Carlsberg Research Laboratory, Department of Chemistry,
Gamle Carlsberg Vej 10, 2500 Valby, Copenhagen, Denmark

U. Sonnewald, Institut für Genbiologische Forschung Berlin GmbH,
Ihnestrasse 63, W-1000 Berlin 33, Germany

B. Svensson, Carlsberg Research Laboratory, Department of Chemistry,
Gamle Carlsberg Vej 10, 2500 Valby, Copenhagen, Denmark

R. Schär, Bühler AG (NM 1), Bahnhofstrasse, 9240 Uzwil, Switzerland

F. Schierbaum, Kantstrasse 4, O-1570 Potsdam, Germany

H.A. Schols, Wageningen Agricultural University, Department of Food Science,
Bomenweg 2, 6703 HD Wageningen, The Netherlands

E.J. Stamhuis, University of Groningen, Department of Chemical Engineering,
Nyenborgh 4, 9747 AG Groningen, The Netherlands

J. Tayeb, INRA, Centre de Recherche Agro-Alimentaires, B.P. 527,
44026 Nantes, France

D. Thorniley, Fairlawn Development Center, Extrusion Technology, Nabisco Brands
Inc., 21-11 Route 208, Fair Lawn, NJ 07410, USA

B. Vergnes, INRA, Centre de Recherche Agro-Alimentaires, B.P. 527, 44026 Nantes, France

A.G.J. Voragen, Wageningen Agricultural University, Department of Food Science, Bomenweg 2, 6703 HD Wageningen, The Netherlands

W. Vorwerg, Fraunhofer-Institute for Applied Polymeric Research Teltow, O-1505 Bergholz-Rehbrücke, Germany

E. Wilhelm, Federal Centre for Cereal, Potato, and Lipid Research, Detmold and Münster, Schützenberg 12, W-4930 Detmold, Germany

L. Willmitzer, Institut für Genbiologische Forschung Berlin GmbH, Ihnestrasse 63, W-1000 Berlin 33, Germany

V.P. Yuriev, INEOS, 125080 Moscow, Russia

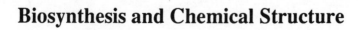

Biosynthesis and Chemical Structure

STRUCTURAL STUDIES OF PLANT CELL-WALL POLYSACCHARIDES USING ENZYMES

A.G.J. Voragen, H.A. Schols and H. Gruppen

DEPARTMENT OF FOOD SCIENCE, WAGENINGEN AGRICULTURAL UNIVERSITY, BOMENWEG 2, 6700 EV WAGENINGEN, THE NETHERLANDS

1   INTRODUCTION

The classical way to establish the chemical structure of a polysaccharide, carefully extracted and purified from its original source, consists of the following steps: 1) establishing its sugar composition, 2) establishing its glycosidic linkage composition by methylation analysis, 3) recognition of repeating units by specific chemical degradation procedures like partial hydrolysis and acetolysis, periodate based oxidation reactions or base-catalysed elimination reactions, 4) NMR-spectrometric methods (for not too complex polysaccharides).

Enzymes have also successfully been used as valuable analytical tools in the elucidation of the fine structure of polysaccharides[1]. The potential use of enzymic fragmentation in the elucidation of the structure of polysaccharides depends largely on the purity of the enzyme used, and on a thorough knowledge of their substrate specificity and pattern of action. The increased availability of such pure, well characterized enzymes as well as recent developments in techniques for separation of enzyme digests of polysaccharides and analytical techniques for revealing the structural features of isolated fragments have greatly enhanced the use of enzymes for polysaccharide structure elucidation.

In this paper we will demonstrate this with results of recent investigations on the structural features of hairy (ramified) regions from apple pectins and of wheat arabinoxylans. In addition, some results of the mass spectrometric identification of oligomeric reaction products present in the endo-(1,4)-ß-D-galactanase digest of a soy arabinogalactan will be shown.

2   EXPERIMENTAL

General Approach

The polysaccharide is first extracted from its source and purified. Enzymes can already be used for the specific extraction of certain polysaccharides as is the case for the isolation of hairy regions of pectins. Homogeneous fractions of the isolated polysaccharide can then be fragmented with pure, well characterized enzymes or combinations of pure enzymes. In the following step the digest containing the polysaccharide fragments is fractionated by size exclusion chromatography and pools of oligomeric fragments in the same molecular weight range are further fractionated in a preparative way by High-Performance Anion-Exchange Chromatography (HPAEC). The structure of the isolated oligomers can then be determined by nuclear magnetic resonance (NMR) spectrometry or mass spectrometry (MS). Finally, the overall structure of the polysaccharide can be constructed from the relative proportions of the structural elements in the parental material, the established structures for the oligosaccharides and the sugar and linkage composition of polymeric fragments in the polysaccharide digest, the substrate specificity and the pattern of action of the enzyme used.

Materials.   Hairy (ramified) regions of pectins. Hairy regions of pectin were isolated from apples and from a number of other plant materials e.g. pears, onions, potatoes, leeks and carrots.For this purpose the materials were ground and enzymically liquefied using a crude mixture of pectinases and cellulases. The (modified) hairy regions (MHR) were recovered from the liquid phase by ultra-filtration and dialysis. Size Exclusion Chromatography (SEC) on Sephacryl S200 and S500 was used to further fractionate MHR. All fractions were characterized for sugar composition, linkage analysis, degree of methylation and acetylation and molecular weight distribution[2]. Degradation studies were performed using the newly isolated enzyme rhamnogalacturonase[3] (RGase). To enhance the enzymic degradation MHR was saponified prior to the enzyme treatment. Enzyme digests were analyzed for changes in molecular weight distribution by High Performance Size Exclusion Chromatography (HPSEC) on three Bio-Gel TSK columns in series (40XL, 30XL, and 20XL) as described[2]. Preparative size exclusion chromatography for isolation of reaction products of different sizes was performed on a Sephacryl S200 column using a 0.1M sodium succinate buffer of pH 4.8 for elution. Oligomeric fragments were isolated by preparative HPAEC. The structure of isolated oligomers was established using ^{1}H and ^{13}C NMR[4].

Wheat Arabinoxylans.   Wheat arabinoxylans were selectively extracted from the water-unextractable cell-wall material[5] of wheat flour with a saturated barium hydroxide solution[6] and subsequently fractionated by DEAE-chromatography. The neutral fraction (BE1-U), representing ca. 80% of the extracted arabinoxylans was subfractionated by graded ethanol precipitation. All fractions were characterized for sugar composition, linkage analysis and molecular weight distribution. This is described in detail by Gruppen et al. [7]

For the fragmentation of the arabinoxylan populations two endo-(1,4)-ß-D-xylanases and an (1,4)-ß-D-arabinoxylan arabinofuranohydrolase, all isolated from the culture medium of *Aspergillus awamori*, were used[8,9]. The fractionation of the arabinoxylan digests in oligomers of similar hydrodynamic volume by size exclusion chromatography on a Bio-Gel P-2 column and the preparative isolation of oligomeric arabinoxylan fragments by HPAEC is described by Gruppen et al.[10] and Kormelink et al.[11] The elucidation of the structure of the isolated oligomers was established by NMR[10,11].

Arabinogalactan.   An arabinogalactan was extracted from the water-unextractable material from dehulled soy beans using 0.1M sodium hydroxide at 0°C. After neutralization, dialysis and lyophilization the extract was fractionated on a DEAE Sepharose Cl-6B column[12]. A fraction enriched in arabinogalactans was fragmented with an endo-(1,4)-ß-D-galactanase. The digest was fractionated in oligomer fractions by size exclusion chromatography using a Bio-Gel P-2 column. The oligosaccharides present in the different Bio-Gel P-2 pools were further analyzed by HPAEC-MS,

Fourier transform ion cyclotron resonance (FTICR)-MS, and Fast Atom Bombardment (FAB)-MS/MS.

3   RESULTS AND DISCUSSION

Hairy (Ramified) Regions of Pectins

Based on studies on the structural features of pectins extracted with a variety of extractants from apple cell-wall material, De Vries et al.[13] proposed a model in which pectin molecules were considered to contain "homogalacturonan" regions comprising 90% of the galacturonan backbone and so-called "hairy" (ramified) regions with a galacturonan backbone rich in rhamnose and carrying the major part of the neutral sugars of the pectins (Figure 1). Using an enzymic liquefaction process Schols et al.[2] isolated pectic substances from apple tissues resistant to further enzymic degradation and having structural features similar to the hairy regions described by De Vries et al.[13] Because of the wide spectrum of carbohydrase activities present in the enzyme preparation used, it must be assumed that alterations of the side chains will have taken place and therefore we designated the isolated material as modified hairy regions (MHR). During recent investigations in our laboratory, we were able to isolate similar pectic material, using the same extraction procedure, from various sources (Table 1). In spite of individual differences between the materials, a great similarity can be observed with respect to rhamnose, galacturonic acid, and acetyl content. Thus it appears that pectic hairy regions are common to the plant material investigated. The occurrence of pectic hairy regions has also been reported for carrots[15] and grape berries[16,17] using different extraction procedures.

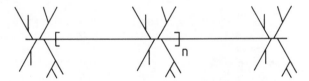

Figure 1   Schematic structure of apple pectin; rhamnogalacturonan backbone with regions rich in neutral sugar side chains (hairy regions) (from Pilnik[14])

The structural features of apple MHR have been characterized in more detail. Using SEC, MHR could be fractionated into three populations (A, B and C) having different hydrodynamic volumes. Although slight differences were found in the three populations (e.g. rhamnose and xylose content, degree of methoxylation and acetylation), the parental MHR can be considered as a mixture of structurally similar polysaccharides. From a commercial enzyme preparation (ex *Aspergillus aculeatus*) a novel enzyme was isolated that could degrade apple MHR[3]. Oligomeric reaction products produced by this enzyme were isolated from a digest and their structures elucidated using NMR-spectroscopy[4]. Based on the observation that these oligomers had a (substituted) rhamnose residue at the non-reducing site and a galacturonic acid residue on the reducing site we named this enzyme Rhamnogalacturonase (RGase). From the sugar and linkage composition of MHR, the sugar composition of fragments obtained by MMNO (4-methyl-morpholino-N-oxide)-catalysed degradation reaction, the resistance to degradation by a variety of pectin and hemicellulose degrading enzymes and from the structures established for the oligomeric degradation products obtained with RGase a tentative structure for MHR was proposed (Figure 2).

Table 1   Sugar composition of pectic polysaccharides (MHR) isolated from the retentate of ultrafiltrated liquefaction juice from carious sources. DM: degree of methoxylation; DA: degree of acetylation (both based on galacturonosyl content)

Sugar	Sugar composition (mol %)					
	Apple	Pear	Leek	Onion	Carrot	Potato
Rhamnose	6	12	22	21	22	16
Arabinose	55	40	18	7	12	16
Xylose	8	7	1	3	1	2
Mannose	0	0	0	1	1	1
Galactose	9	14	24	30	24	24
Glucose	1	0	0	0	1	6
Uronic acid	21	27	35	38	39	35
DM	42	21	13	6	4	12
DA	60	33	77	61	55	90

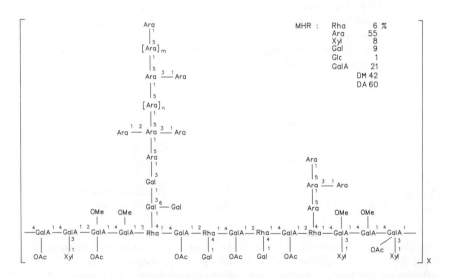

Figure 2   Hypothetical structure of the modified hairy regions of apple (from Schols et al.[2])

The study of the degradation products formed from MHR was hampered by the heterogeneity of MHR. Degradation products of the high molecular weight (Mw) population A co-eluted with non-degraded parts of the populations B and C.

In the continuation of our studies we therefore used only population A, which is more homogeneous and represents the major part of the polysaccharides in MHR. Fraction A was saponified with alkali and incubated with RGase. The degradation was monitored by HPSEC. The elution profiles after various reaction times are shown in Figure 3. A rapid shift in the molecular weight distribution, indicating an endo-type of action of RGase can be seen.

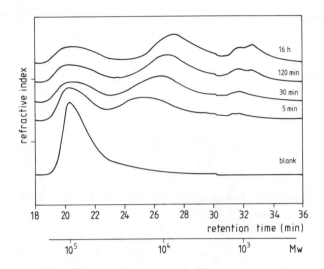

<u>Figure 3</u>   High performance size exclusion chromatography of MHR population A, before and after treatment with RGase at 30°C and pH 5.0

Oligomeric reaction products were formed in later stages of the reaction by degradation of the "early state" reaction products. To enable structural studies of the reaction products the digest remaining after exhaustive degradation with RGase was fractionated by SEC on a Sephacryl S200 column. From the elution profile in Figure 4 three populations of degradation products can be distinguished. They were pooled in four fractions as indicated in the figure.

The sugar composition of the parental fraction A and of the four pools are given in Figure 4. It can be seen that xylose and galacturonic acid (ratio 1:1) accounted for about 80% of the constituent sugars in the void fraction I. This is in sharp contrast with the starting material in which arabinose prevails, while xylose and galacturonic acid only make up 35% of the constituent sugars. These observations indicate the presence of xylogalacturonan segments in the modified hairy regions. Xylogalacturonans, isolated from the pollen of the Mountain Pine, have already been described by Bouveng[18].

Fractions II and III covered a broad Mw range and consisted both of arabinose-rich polysaccharides (> 80% arabinose). Apparently these fractions represent arabinans, connected to some residual stubs of the pectic backbone.

Fraction IV represented oligomeric fragments and corresponded with material eluting at retention times of 31-33 min (Figure 3). From the sugar compositions, it can also be calculated that the rhamnose/galacturonic acid ratio was highest in the smaller fragments (pool IV).

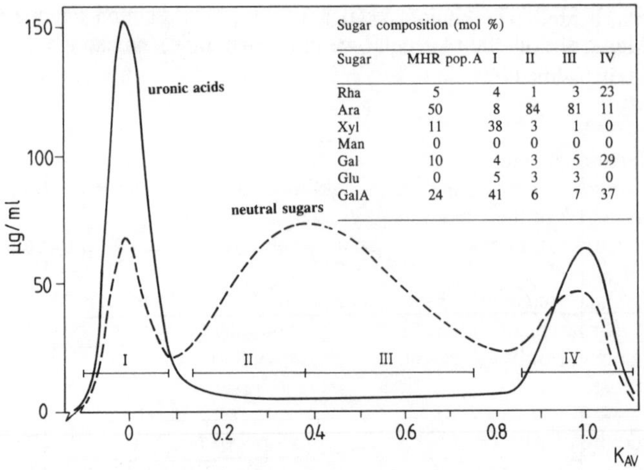

Sugar composition (mol %)

Sugar	MHR pop.A	I	II	III	IV
Rha	5	4	1	3	23
Ara	50	8	84	81	11
Xyl	11	38	3	1	0
Man	0	0	0	0	0
Gal	10	4	3	5	29
Glu	0	5	3	3	0
GalA	24	41	6	7	37

Figure 4 Size exclusion chromatography on Sephacryl S200 of MHR population A after degradation by RGase

The availability of the strong anion-exchange resin CarboPac PA1 in combination with pulsed amperometric detection enabled us to separate a range of oligomers. Using a semi-preparative column (9•250 mm ID), these oligomers were isolated and characterized by NMR. They were found to have a backbone of alternating rhamnose and galacturonic acid residues with rhamnose at the non-reducing end, some of the oligomers carry galactosyl residues on all or part of the rhamnose residues[4]. Figure 5 shows a typical chromatogram of a RGase digest of fraction A. In the digest we can recognize the tetramer, the two pentamers and the hexamers already identified by ^{13}C-NMR[4]. In addition larger oligomers with basically the same structural features eluting at higher retention times were identified.

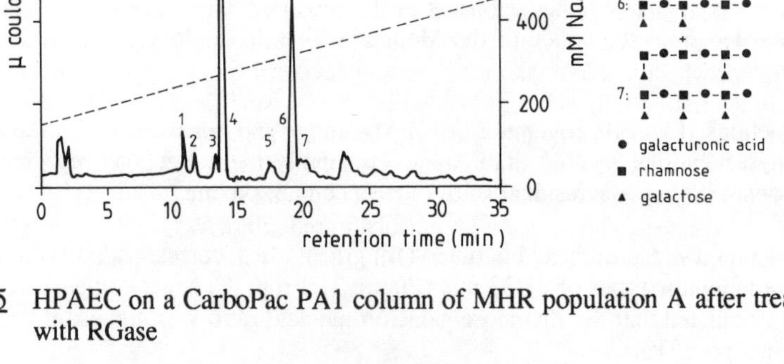

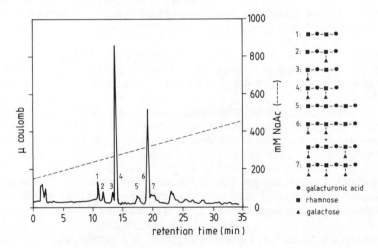

Figure 5 HPAEC on a CarboPac PA1 column of MHR population A after treatment with RGase

Pectic hairy regions appear as rather complex polysaccharides. From the presented characteristics of the degradation products of the more homogeneous fraction A of apple MHR formed by RGase action at least three different repeating units can be distinguished:

I xylogalacturonan units consisting of an (1,4)-α-D-galacturonan backbone with some rhamnose inserted and carrying single unit side chains of xylosyl residues ß-linked to O-3 of the galacturonosyl residue.

II residual stubs of the pectic backbone containing arabinan side chains (> 80% arabinose).

III rhamnogalacturonan chains of unknown length, consisting of alternating (1,2)-linked α-L-rhamnose and (1,4)-linked α-D-galacturonic acid residues with part of the rhamnosyl residues carrying single unit galactose side chains, linked to the O-4 of rhamnose.

The structure of MHR has been revised in view of these new data and is shown in Figure 6. It should be considered that the order of the different subunits in the model is purely hypothetical and is subject to our further research.

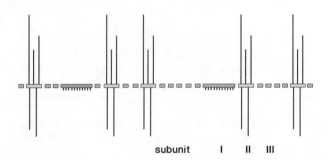

subunit   I   II   III

<u>Figure 6</u>   Schematic structure of apple modified hairy regions. Subunit I, xylogalacturonan; subunit II, Arabinans linked to rhamnogalacturonan; subunit III, rhamnogalacturonan oligomers

Wheat Arabinoxylans

For the characterization of hairy regions of pectins, we started with enzymically extracted, rather heterogeneous and structurally altered pectic material. To establish the structure of native polysaccharides, homogeneous intact starting material is a requisite. For wheat arabinoxylans, this was achieved by preparing water unextractable cell-wall material (WUS) under mild conditions[5,19]. Using the procedures described in the experimental section a highly purified, neutral arabinoxylan preparation (BE1-U) accounting for about 65% of the arabinoxylan present in wheat flour WUS was obtained. From the sugar composition and glycosidic linkage analysis it could be derived that 64% of the xylosyl residues were unsubstituted, 20% was single substituted with arabinosyl units either at O-2 (2%) or at O-3 (18%), 16% was double substituted with arabinosyl units at O-2,3.

   Two types of pure and well characterized endo-(1,4)-ß-D-xylanases, designated
as endo I and endo III, were used to degrade the arabinoxylan. This degradation was
carried out with the xylanases alone or in combination with a (1,4)-ß-D-arabinoxylan
arabinofuranohydrolase (AXH). The latter enzyme is able to remove O-3 linked
arabinofuranosyl residues from single substituted xylopyranosyl units only[20]. Figure 7
shows the oligosaccharide distribution of the arabinoxylan digests with endo I and
endo III as obtained with Bio-Gel P-2 size exclusion chromatography. The numbers 1
to 10 correspond with pools of oligosaccharides with degree of polymerization of 1 to
10, respectively. The elution patterns clearly show differences in the pattern of action
of both endo-xylanases. Pools of oligomers with the same size were further fractionated by preparative HPAEC and most oligomer mixtures could be resolved to pure
oligosaccharides (Figure 8). The structures of these isolated oligosaccharides were
elucidated by NMR-spectroscopy[10,11]. Some typical identified arabinoxylan oligosaccharides are shown in Figure 9.

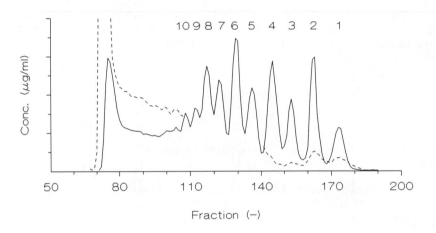

<u>Figure 7</u>   Elution profile of both endo I (—) and endo III (---) arabinoxylan digest on
          Bio-Gel P-2

   The oligomers from endo I had one unsubstituted xylosyl residue at the reducing
side when this residue was adjacent to an O-3 substituted xylosyl residue and two unsubstituted xylosyl residues when they were adjacent to an O-2,3 substituted residue.
The oligomers of endo III all had 2 unsubstituted xylosyl residues at their reducing
side. With respect to the nature of the non-reducing terminal sugar unit 2 groups of
arabinoxylan-oligosaccharides can be distinguished in each endo-xylanase digest. For
endo I these are: A) oligosaccharides with a substituted xylose residue at the non-reducing end and B) oligosaccharides with 1 unsubstituted xylosyl residue adjacent to a
substituted xylosyl residue at the non-reducing end. For endo III these are: C) oligosaccharides with 1 unsubstituted xylose residue and D) oligosaccharides with 2 unsubstituted xylosyl residues adjacent to a substituted xylosyl residue at the non-reducing end.

   About 10% of the arabinoxylan remained as polymeric material (void fraction in
Figure 7) after degradation with endo I and was resistant to further attack. For endo
III this fraction accounted for 20% of the arabinoxylan.

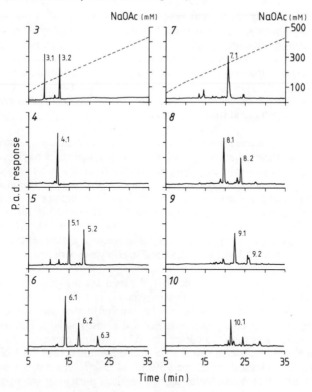

Figure 8 Elution profile on HPAEC of Bio-Gel P-2 fractions 3-10 from endo I digest in Figure 7 (from Gruppen et al.[10])

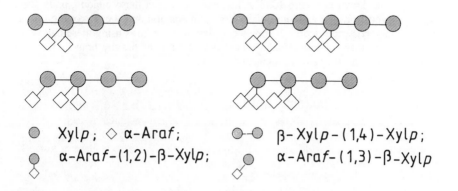

Figure 9 Typical structure of endo I derived oligosaccharides

Pretreatment of the arabinoxylan with AXH prior to incubation with endo I or endo III resulted in the presence of 11.5% monomeric arabinose, expressed as percentage of total carbohydrate material present in the parental material, almost equal to the theoretical value based on arabinose substitution at O-3. The amounts of xylose and higher xylo-oligomers present in both the endo I as well as in the endo III digest increased upon pretreatment with AXH. Arabinoxylan-oligomers containing O-3 linked arabinose were absent after AXH treatment. The polymeric enzyme resistant fraction after endo I treatment decreased from 10 to ca. 6% and of endo III from 20 to 14%. The arabinose/xylose (Ara/Xyl) ratio of the endo I resistant polymeric fraction was found to decrease upon pretreatment with AXH, whereas the Ara/Xyl ratio of the endo III resistant polymeric fraction increased.

From the relative abundance of identified oligosaccharides, the described mode of action of the enzymes[21], the linkage analysis of the parental arabinoxylan, and the composition of the enzyme-resistant polymeric material remaining after enzyme action, information on the structure of the arabinoxylan can be deduced. The void fraction of the endo I digest and the amounts, and structures of the oligomeric reaction products in the endo III digest are of special importance in this respect. In the endo III digest about 49% of the arabinoxylan was present as oligomers with a DP≤10 with many oligosaccharides having either one O-3 or O-2,3 substituted xylose residue and oligosaccharides with both an O-3 and an O-2,3 substituted xylose residues.

From the mode of action of endo III[21], the total amount of the two groups of oligosaccharides described above for this enzyme (C and D) and the presence of unsubstituted xylo-oligosaccharides, it can be concluded that the native arabinoxylan must contain regions of contiguous unsubstituted xylose residues. The endo I resistant polymeric fraction was found to contain 8% of the xylose present in the parental material. It can also be calculated that this polymeric fraction contains 5% of all unsubstituted, 5% of all O-3 substituted, and 25% of all O-2,3 substituted xylosyl residues. The low amount of O-2 substituted xylosyl residues present in the parental material does not allow an estimation of the distribution of this building unit.

From the mode of action of endo I it can be concluded that the polymeric fraction remaining after incubation of the arabinoxylan with endo I must be built with the building units shown in Figure 10. The frequency in which these building units occur can be derived from the arabinose and xylose composition and the glycosidic linkage types of the polymeric fraction. Involved calculation procedures using these data allow the conclusion that arabinoxylans in wheat flour consist of heavily branched regions interlinked with linear or slightly branched regions.

building units:

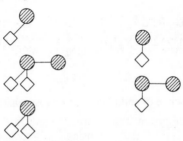

Figure 10 Buildings units of polymeric fraction in endo I digest

By graded ethanol precipitation the neutral arabinoxylan fraction could be sub-fractionated into five major fractions precipitating successively at 20, 30, 40, 50 and 60% ethanol. These fractions had Ara/Xyl ratios increasing from 0.36 to 0.80. With increasing Ara/Xyl ratio the molar proportion of unsubstituted and O-3 substituted xylosyl residues were found to decrease while the molar proportion of O-2,3 substituted xylosyl residues increased strongly and the molar proportion of O-2 only slightly. This is shown in Figure 11. The effect of the Ara/Xyl ratio of the parental arabinoxylan on the composition of the digests obtained with endo I or endo III is shown in Figure 12. With increasing Ara/Xyl ratio the concentration of xylose and low oligomeric reaction products rapidly decrease while the concentration of larger (DP > 6) oligomers increase to Ara/Xyl ratio 0.68 and then decrease. In addition, the polymeric void fraction ('13' in Figure 12) rapidly increases when arabinoxylans with increasing Ara/Xyl ratio are used as substrate. Upon analysis of the different Bio-Gel P-2 fractions with HPAEC it was observed that with increasing Ara/Xyl ratio of the parental material in general decreasing amounts of oligosaccharides containing O-3 substituted xylosyl residues and increasing amounts of oligosaccharides containing O-2,3 double substituted xylosyl residues were found. The amount of oligosaccharides containing both O-3 and O-2,3 substituted xylosyl residues showed little variation with Ara/Xyl ratio.

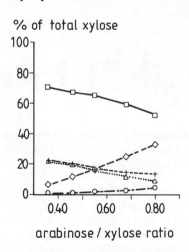

Figure 11 Relative amounts of partially methylated xylose residues in BE1-U subfractions; □ 2,3-Me$_2$-Xyl; + 2(3)-Me-Xyl; ◊ Xyl; △ 2-Me-Xyl; O 3-Me-Xyl (from Gruppen et al.[7])

These results indicate that arabinoxylans with different Ara/Xyl ratios vary both in the relative amount as well as in the composition of structural elements of the less branched regions.

New Developments

In the examples described above structure elucidation of oligomeric reaction products formed by enzymic degradation using NMR-spectroscopy is an essential step. However, other analytical techniques like mass spectrometry are gaining more interest.

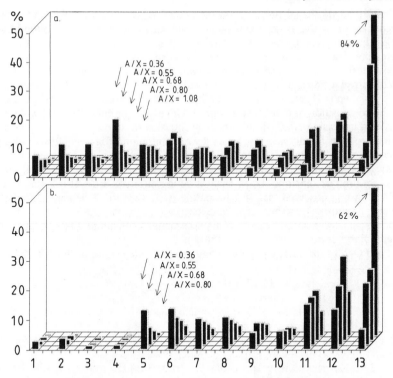

<u>Figure 12</u> Distribution of oligomers (1-12) and polymeric material (13) on Bio-Gel P-2 from endoxylanase I (a) and endoxylanase III (b) digests of different alkali-extractable arabinoxylans

Niessen et al.[22,23] were successful in combining HPAEC with a mass spectrometer using a thermospray interface. Using this system, they were able to analyze a digest of a soy arabinogalactan obtained by degradation with an endo-galactanase and Mw information was obtained for the different oligomers present. Since there is no clear regularity in the elution order of the oligomers, which depends on size, sugar residue and linkage types present[24], Mw information as obtained by HPAEC-MS is very helpful. In combination with other available information like extraction conditions of the polysaccharide, sugar composition of the polymer, enzyme used for the degradation etc., the Mw of the various oligomers is often sufficient for revealing the structure of the oligomers.

Research at the Institute of Mass Spectrometry of the University of Amsterdam (Dr. N.M.M. Nibbering and co-workers), using Fourier transform ion cyclotron resonance MS techniques, revealed Mw information for the various oligomers present in the "nonamer"-fraction, isolated by size exclusion chromatography on Bio-Gel P-2. Tandem (MS/MS) mass spectrometry under Fast Atom Bombardment conditions revealed the structure of two different nonamers in the mixture. One being a linear galactose oligomer, while the other contained two single unit arabinose side groups. Also information on the location of the arabinose groups was obtained.

These examples demonstrate that these hyphenated techniques have the potential to facilitate the characterization of mixtures of oligomeric fragments omitting labour-intensive preparative isolation procedures.

4 REFERENCES

1. N.K. Matheson and B.V. McCleary, 'The Polysaccharides', (Ed. G.O. Aspinall), Academic Press, 1985, Vol. 3, 1.
2. H.A.Schols, M.A. Posthumus and A.G.J. Voragen, Carbohydr. Res., 1990, 206, 117.
3. H.A. Schols, C.C.J.M. Geraeds, M.J.F. Searle-van Leeuwen, F.J.M. Kormelink and A.G.J. Voragen, Carbohydr. Res., 1990, 206, 105.
4. I.J. Colquhoun, G.A. De Ruiter, H.A. Schols and A.G.J. Voragen, Carbohydr. Res., 1990, 206, 131.
5. H. Gruppen, J.P. Marseille, A.G.J Voragen and R.J. Hamer, Cereal Chem., 1990, 67, 512.
6. H. Gruppen, R.J. Hamer and A.G.J. Voragen, J. Cereal Sci., 1991, 13, 275.
7. H. Gruppen, R.J. Hamer and A.G.J. Voragen, J. Cereal Sci., 1992, 16, 53.
8. F.J.M. Kormelink, M.J.F. Searle-van Leeuwen, T.M. Wood and A.G.J. Voragen, J. of Biotechnol., in press.
9. F.J.M. Kormelink, M.J.F. Searle-van Leeuwen, T.M. Wood and A.G.J. Voragen, Appl. Microb. Biotechnol., 1991, 35, 753.
10. H. Gruppen, R.A. Hoffmann, F.J.M. Kormelink, A.G.J. Voragen, J.P. Kamerling and J.F.G. Vliegenthart, Carbohydr. Res., 1992, 233, 45.
11. F.J.M. Kormelink, R.A. Hoffmann, H. Gruppen, A.G.J. Voragen, J.P. Kamerling and J.F.G. Vliegenthart, unpublished results.
12. H.A. Schols, G. Lucas-Lokhorst and A.G.J. Voragen, 'Publications from the Department of Cereal Technology, 11th issue', Technical University of Berlin, Germany, in preparation.
13. J.A. De Vries, F.M. Rombouts, A.G.J. Voragen and W. Pilnik, Carbohydr. Polym., 1982, 2, 25.
14. W. Pilnik, 'Uses of Enzymes in Food Technology', (Ed. P. Dupuy), Technique et Documentation Lavoisier, 1982, p. 425.
15. H. Konno, Y. Yamasaki and K. Katoh, Phytochemistry, 1986, 25, 623.
16. L.Saulnier and J. F. Thibault, Carbohydr. Polym., 1987, 7, 345.
17. L.Saulnier, J.-M. Brillouet and J.-P. Joseleau, Carbohydr. Res., 1988, 182, 63.
18. H.A. Bouveng, Acta Chem. Scand., 1965, 19, pp. 953.
19. H. Gruppen, J.P. Marseille, A.G.J Voragen, R.J. Hamer and W. Pilnik, J. Cereal Sci., 1989, 9, 247.
20. F.J.M. Kormelink, M.J.F. Searle-van Leeuwen, T.M. Wood and A.G.J. Voragen, Appl. Microb. Biotechnol., 1991, 35, 231.
21. F.J.M. Kormelink, H. Gruppen, R.J. Viëtor and A.G.J. Voragen, unpublished results.
22. W.M.A. Niessen, R.A.M. van der Hoeven, J. van der Greef, H.A. Schols, G. Lucas-Lokhorst, A.G.J. Voragen and C. Bruggink, Rapid Commun. Mass Spectrom., 1992, 6, 474.
23. W.M.A. Niessen, R.A.M. van der Hoeven, H.A. Schols, G. Lucas-Lokhorst, J. van der Greef, A.G.J. Voragen and C. Bruggink, 'Publications from the Department of Cereal Technology, 11th issue', Technical University of Berlin, Germany, in preparation.
24. Y.C. Lee, Anal. Biochem., 1990, 189, 151.

A NEW METHOD OF ENZYMIC ANALYSIS OF AMYLOPECTIN STRUCTURE

S. Hizukuri and J. Abe

DEPARTMENT OF BIOCHEMICAL SCIENCE AND TECHNOLOGY, KAGOSHIMA UNIVERSITY, KORIMOTO-1, KAGOSHIMA 890, JAPAN

1 INTRODUCTION

Starches from various plant origins exhibit characteristic functional properties depending on the integrated factors molecular structure, higher structures including crystalline state, and minor constituents such as lipids, phosphate, and proteins Among these factors, molecular structure is of primary importance, since it even directs the crystalline structures of starch granules[1-4]. The basic structures of amylose and amylopectin have been elucidated during several decades. However, their details, especially of branched molecules, have not been well understood. To elucidate the structures of branched molecules, we have to analyze the sizes and amounts of individual chains and the mode of connections of the chains. In this study, the latter has been focused on.

Amylopectin is composed of A, B, and C chains: that is linear A chains, which link to B or C chains by their terminal reducing residues; branched B chains, which carry the A and/or B chain(s) at their C-6; and the C chain having a reducing residue[5]. Amylopectin is built up with equal numbers of A and B chains or a small excess of A chains[6,7] and only one C chain.

Recently, Hizukuri and Maehara[8,9] have found that some B chains of waxy rice and wheat amylopectin carried an increased number of A chains with increased chain lengths, and some B chains (Bb chain) carried no A chains, by successive degradation with β-amylase, isoamylase and β-amylase, i.e. β, i, β-degradation. This technique gives quantitative information on the linking between chains but a correction is necessary for the partial liberation of maltosyl residues derived from the A chains and there is a little difficulty in finding proper conditions for the isoamylolysis of β-limit dextrin. Now we report a successive degradation with glucoamylase and isoamylase, i.e. γ, i-degradation, which gives more informative results on the mode of chain-linking.

2 THEORY

The modes of chain connections of amylopectin or glycogen can be classified into Staudinger(S) and Haworth(H) types (Figure 1). In the H type, one chain is linked by its reducing residue to C-6 of the other chain, which is linked to the third chain also by own reducing residue; in the S type, two chains are linked to another single chain by their reducing residues. Staudinger and Haworth structures of amylopectin comprise solely S and H type connections, respectively. However, the Meyer irregular structure is a mixture of the S and H types.

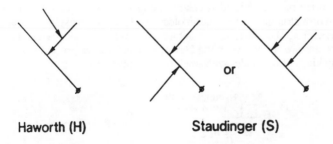

Haworth (H)   **Staudinger (S)**

<u>Figure 1</u>   Two types of chain connections Haworth (H) and Staudinger (S) types

<u>Table 1</u>   Amylopectin structures and their γ, i-degradation products (γ, i-LD)

Structure	Location of phosphate	γ,i-LD	
		Phosphorylated	Neutral
Staudinger	A chain	$\overset{P}{\rule{0.6cm}{0.4pt}}\!\!\to\varnothing$ (S*)	$\underset{n\geq1}{\rule{0.6cm}{0.4pt}}\!\!\to\varnothing$ (L*)
	C chain	$\underset{n\geq1}{\overset{P}{\rule{0.6cm}{0.4pt}}}\!\!\to\varnothing$ (L)	
Haworth	A chain	$\overset{P}{\rule{0.6cm}{0.4pt}}\!\!\to\varnothing$ (S)	$\rule{0.6cm}{0.4pt}\!\!\to\varnothing$ (S)
	B chain	$\underset{n=0,1}{\overset{P}{\rule{0.6cm}{0.4pt}}}\!\!\to\varnothing$ (S)	
Meyer or cluster	A chain	$\overset{P}{\rule{0.6cm}{0.4pt}}\!\!\to\varnothing$ (S)	$\rule{0.6cm}{0.4pt}\!\!\to\varnothing$ (S)
	B chain	$\underset{n\geq0}{\overset{P}{\rule{0.6cm}{0.4pt}}}\!\!\to\varnothing$ (S,M*)	$\underset{n\geq0}{\rule{0.6cm}{0.4pt}}\!\!\to\varnothing$ (S,M)

*S, M, and L, small, medium, and large molecules; ♀ , α-glucosyl branch; $\varnothing$, reducing residue, —, linear chain

Amylopectin contains a small proportion of glucose residues with phosphate ester groups at C-6 and C-3[10-12]. These phosphate groups are located mainly in the outer section of the B chains[13]. Glucoamylase hydrolyzes amylopectin from non-reducing residues of the chains up to a glucose 6-phosphate residue or to one residue on the outer side of a glucose 3-phosphate residue[14-16]. When A or B chains are linked by their reducing residues to C-6 of a phosphorylated B chain at the inner side of the phosphate, these side chains are trimmed by glucoamylase and remain as $(1\to6)$-α-glucosyl stubs. When a phosphorylated chain (P-chain) carries n neutral chains (N-chain) at the outer or the inner part of the location of phosphate, it yields a linear P-chain fragment or a $[(1\to6)$-α-glucosyl$]_n$-P-chain fragment by successive actions of glucoamylase and isoamylase, respectively. When a P-chain is linked by its reducing residue to an N-chain with n side chains, the N-chain yields a $[(1\to6)$-α-glucosyl$]$

n'-chain fragment (0≤n'≤n) by the successive actions of the two enzymes. These relationships between the modes of branching and the structures of the γ, i-degradation products (γ, i-LD) are shown in Table 1 and Figure 2. Linear P-chains are formed mainly from the phosphorylated A chains and also from the phosphorylated B chains of the Haworth structure. The larger molecules of multi-glucosylated P- and N-chains are produced from the Staudinger structure, whereas mono-glucosylated P-chains and linear N-chains are formed from the Haworth structure. Meyer and cluster structures yield many kinds of molecules with small or moderate sizes.

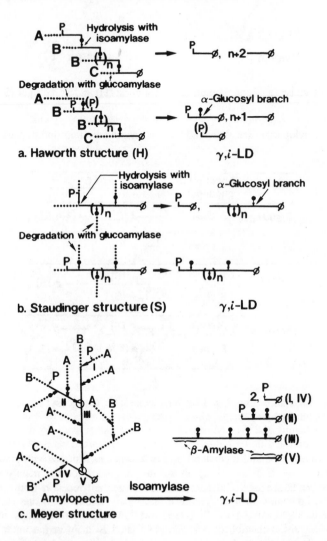

Figure 2 Degradation of some amylopectin structures with glucoamylase, isoamylase, and β-amylase —, (1→4)-α-Glucan chain; ♥, (1→6)-α-glucosyl stub; P, phosphate ester; Ø, reducing residue; →, (1→6)-α-linkage; O, B→B chain linkage hydrolyzed with isoamylase

The average span length between the branches of the γ, i-LD can be calculated from the average main chain length and the number of $(1{\rightarrow}6)$-α-glucosyl stubs. The main chain length can be determined by subtraction of the number of the glucosyl stubs from the d.p.n of the molecule. Some branch linkages between the B chains, which were located in the inner part of the γ-LD, were removed by the isoamylase as shown in Figure 2. Therefore, to calculate the span length of amylopectin, the error due to the removal of the B-B chain linkages should be corrected. This can be done partly by removal of the outer linear parts of γ, i-LD with β-amylase (Figure 2).

3 MATERIALS AND METHODS

Materials

Potato tubers, variety Benimaru, were donated from the Hokkaido Agricultural Experimental Station. The tubers were peeled, sliced in small pieces, and then homogenized with ice-cold water in a home blender. The homogenate was squeezed through cotton cloth. Starch was recovered from the starch suspension by natural sedimentation and decantation. The starch was washed repeatedly with distilled water by sedimentation and decantation, and dried at room temperature. Amylopectin was separated from the starch by modification[17] of the method of Lansky et al.[18] without autoclaving under N_2 atmosphere.

Glucoamylase (*Rhizopus delemar*, GIII[16]) was prepared as described previously. The enzyme was extensively purified to be free from α-amylase by a sensitive assay with using hydroxy-propylated corn starch. Isoamylase (*Pseudomonas amyloderamosa*, crystalline) was a product of Hayashibara Biochemical Lab. Inc. (Okayama, Japan). Glycerol kinase and glycerol phosphate dehydrogenase, and alkaline phosphatase (*Escherichia coli*, type III), which were used for the assay of glycerol produced by Smith degradation, were purchased from Boehringer-Mannheim-Yamanouchi and Sigma Chem. Co., respectively. Other chemicals and reagents were the highest grade available commercially.

Preparation of P- and N-Chains of Amylopectin. Amylopectin (920 mg) was debranched with isoamylase (0.03 U/mg) in 50 mM acetate buffer (pH 3.5) at 50°C for 2.5 h. The hydrolyzate was put on a DEAE-Sephadex A-50 column, and the column was washed with water until no carbohydrate was detected. N-chains were recovered from the effluent and washings. Then, P-chains were eluted from the column with 0.01 M HCl containing 0.2 M sodium chloride. The eluate was neutralized with 1 M NaOH, concentrated under reduced pressure, and deionized by Bio-Gel P-2 gel filtration. The yields of N- and P-chains were 780 and 120 mg, respectively.

Preparations of γ, i-LD and its Fractionation. Amylopectin (5 g) was digested with purified *Rhizopus delemar* glucoamylase GIII as described previously[16]. The yield of γ-LD was about 10%. γ-LD (350 mg) was hydrolyzed with 10.5 IU isoamylase in 175 ml of 50 mM acetate buffer, pH 3.5, at 45°C for 12 h. After heating the mixture at 100°C for 10 min, the mixture was concentrated to 20 ml by a rotary evaporator (γ, i-LD). The concentrate (5 ml) was fractionated by size-exclusion chromatography by passing through a Bio-Gel P-2 column (3.2•89 cm) and each 5 ml fraction was collected. The aliquot (10 ml) of the concentrate of γ, i-LD was put on a column packed with DEAE-Sephadex A-50 (Cl⁻, 100 ml), and washed with distilled water until negative to the anthrone test. The effluent and washings were combined (N-chain), and the column was eluted with 0.01 M HCl containing 200 mM sodium chloride, until nega-tive to the anthrone test. The effluent (N-chain) and the neutralized eluate (P-chain) with 1 M NaOH were concentrated and desalted by means of a Bio-Gel P-2 column.

Analytical Methods

The various analytical methods used are described in the literature cited[10,13].

4   RESULTS AND DISCUSSION

The potato amylopectin used in this study contained 667 ppm of organic phosphorus, of which 64% was located at C-6. Its number-average chain length (c.l.n) was 23.0, the β-amylolysis limit was 56%, and the inner chain length was 7.6. It was composed of 7.6% (by mole) of P-chains with c.l.n 42.2, which carried one phosphate group, and 92.4% N-chains with c.l.n 22.7 (Table 2). This is consistent with previous re-sults[13] that the phosphate ester groups are located mainly in the B chains. The amy-lopectin was hydrolyzed into glucose (86.4% at the limit) with glucoamylase (*Rhizopus delemar*, GIII). The residual dextrin (γ-LD) was hydrolyzed with isoamy-lase into chains (γ, *i*-LD) which carried (1→6)-α-glucosyl stubs. The γ, *i*-LD had d.p.n 29.6, c.l.n 14.9, and one glucosyl stub on average and comprised 60% by weight (50% by mole) of P-chains with d.p.n 37.1 and 40% by weight (50% by mole) of N-chains with d.p.n 24.8. The fractionation of γ, *i*-LD by Bio-Gel P-2 (Figure 3) showed that its d.p.n distributed between 17.0 and 47.9, and it had 1.5 to 3.4 non-re-ducing residues. This means that they had 0-3 glucosyl stubs per main chain and the larger molecules carried an increased number of glucosyl residues derived from side chains. The c.l.n of the main chains was in the range of 16.5 to 45.5. The average span length (s.l.n) (Table 3) suggests that on average, they had branches mostly at every 11 to 12 glucosyl residues. From the distributions of carbohydrate and phosphorus, the P- and N-chains seemed to have similar distributions, and this was supported by the frac-tionation of the separated P- and N-chains, although the P-chains of γ, *i*-LD had a slightly sharper distribution (Figure 4 and Table 4) than that of the N-chains (Figure 5 and Table 5). These results suggest that the skeletal or inner core structure of the po-tato amylopectin is mainly composed of B chains of d.p.n 30 or more, on average having 1-4 A or B chains, including the terminal core B chain, by S type connection as shown in Figure 6. This is similar to the elongated Whelan structure containing the S and H types of chain connections[6,19,20], but some B chains link to other B chains by S type and form two- or three-dimensional structures.

Table 2   Properties of phosphorylated (P) and neutral (N) chains of potato[a] amy-lopectin

Property	Amylopectin	P-chain	N-chain
C.l.n.	23.0	42.2	22.7
D.p.n/Po	287	41.4	-
Amount:			
By weight (%)	100	13.3	86.7
By mole (%)	100	7.6	92.4
Ratio (by mole)		1 : 12	

[a]Benimaru

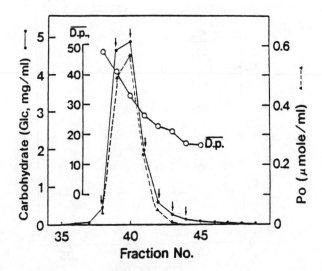

Figure 3   Chromatogram of γ, *i*-LD of potato amylopectin on Bio-Gel P-2:
sample size, 90 mg/5ml; column size, 3.2•89 cm; eluate, H_2O; flow rate, 26
ml/h; fraction size, 5 ml; →, fractions listed in Table 3

Table 3   Analytical data of γ, *i*-LD fractions[a]

Property	Fraction No.							Average
	38	39	40	41	42	43	44	
D.p.n.	47.9	41.3	33.2	26.5	22.8	21.1	17.0	30.0
No. of Glc stub (A)	2.4	2.2	1.7	1.2	0.93	0.99	0.49	1.42
C.l.n. of C chain (B)	45.5	39.1	31.5	25.3	21.9	20.1	16.5	28.6
Span length (S.l.[b])	13.4	12.2	11.7	11.5	11.3	10.1	11.1	11.6

[a]See Figure 3   [b]Average span length = B/(A+1)

The average glucosyl span lengths (s.l.n) between glucosyl stubs of γ, *i-LD* were
found to be in the range of 10-13, but the real s.l.n in the amylopectin is possibly a little shorter than the values because some linkages between two B chains were hydrolyzed with isoamylase as shown in Figure 2 (c. Meyer structure). Therefore, the s.l.n
between branch linkages in amylopectin should be corrected for this factor. This can
be done partly by hydrolysis with β-amylase, which removes the outer linear parts
formed by the debranching of the B-B chain linkage as shown in Figure 2. However,
the β-amylolysis also removes the linear molecules, which were produced by the debranching, so that the correction cannot be done simply by the limit of the hydrolysis.

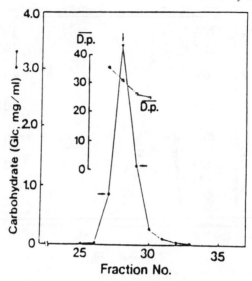

Figure 4   Chromatogram of the P-chains of γ, *i*-LD on Bio-Gel P-2: conditions, see Figure 3; →, fractions listed in Table 4

Table 4   Analytical data of the P-chain fractions[a] of γ, i-LD

	Fraction No.			
	27	28	29	
Property				Average
$\overline{D.p.n.}$	35.7	31.0	26.0	30.9
No. of Glc stub (A)	2.0	1.4	1.0	1.5
C.l.n of C chain (B)	33.7	29.6	25.0	29.4
Span length ($\overline{S.l.}$[b])	11.2	12.3	12.5	12.0

[a]See Figure 4   [b]$\overline{S.l.}$ = B/ (A+l)

The number of branch linkages in branched molecules should be analyzed after removal of maltose from the hydrolysate. However, the large molecular fractions (Table 5, fraction 29 and 30) may contain a negligible amount of linear chains and may give a reasonable value by a simple correction of β-amylolysis limit. This corrected c.l.was about 10. The value is a little higher than the value of the average inner chain length plus 1 (branched residue) of 8.1. This slightly smaller s.l.n than the average inner chain length is due to the partial correction for the removal of the B-B chain linkages and the small amounts of unanalysed fractions which were derived from the heavily branched parts of the molecules. Therefore, the s.l. of the amylopectin would be a little less than 10 and is possibly about 9. The present results provide evidence that the long B chains carried A or B chains at somewhat regular intervals. Evidently,

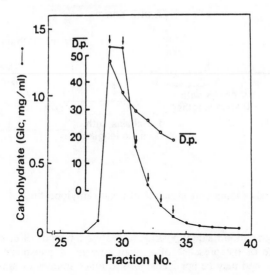

Figure 5   Chromatogram of the N-chains of γ, *i*-LD on Bio-Gel P-2: conditions, see Figure 3; →, fractions listed in Table 5

Table 5   Analytical data of the N-chain fractions[a] of γ, *i*-LD

	Fraction No.						
Property	29	30	31	32	33	34	Average
$\overline{\text{D.p.n.}}$	47.6	35.8	29.1	26.1	21.4	18.6	29.8
No. of Glc stub (A)	2.3	1.8	1.2	0.9	0.5	0.7	1.2
C.l.n. of C chain (B)	45.3	34.0	27.9	25.2	20.9	17.9	28.5
Span length (S.l.[b])	13.7	12.1	12.7	13.2	13.9	10.5	12.7
β-Limit (%)	18.5	25.0	32.8	38.8	34.8	41.1	
$\overline{\text{Corrected}}$ S.l.[c]	11.1	9.1					

[a] See Figure 5   [b] $\overline{\text{S.l.}} = \text{B}/(\text{A}+1)$ [c] Assuming no linear chains, see text

some small γ, *i*-LD molecules, from the numbers of glucosyl stubs, were linear. It implies the presence of some Bb chains, which have no A chains.

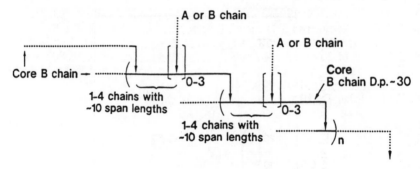

Figure 6   Skeletal or inner core structure of potato amylopectin

The γ, i-degradation made it possible to analyse the mode of chain connection by taking advantage of the presence of a small amount of phosphate in potato amylopectin. The method may be applicable also to other sources of amylopectins which bind by esterification no or extremely small amounts of phosphate, such as wheat and some cereal amylopectins and glycogens. For this a small amount of phosphate can be incorporated, which impedes the progress of glucoamylase action.

5   CONCLUSIONS

Branching mode of potato amylopectin comprises a mixture of S and H type structures, thus forming the Meyer type irregular structure. Some B chains of potato amylopectin carry 1-4 A or B chains by the S type connection and some B chains do not carry A chains. The span lengths between the branch linkages are suggested to be on average about 10.

6   REFERENCES

1. S. Hizukuri, T. Kaneko and Y. Takeda, <u>Biochim. Biophys. Acta</u>, 1983, <u>760</u>, 188.
2. S. Hizukuri, <u>Carbohydr. Res.</u>, 1985, <u>141</u>, 295.
3. B. Pfannemüller, <u>Int. J. Biol. Macromol.</u>, 1987, <u>9</u>, 105.
4. M. Gidley, <u>Carbohydr. Res.</u>, 1987, <u>161</u>, 301.
5. S. Peat, W.J. Whelan and G.J. Thomas, <u>J. Chem. Soc.</u>, 1952, 4546.
6. D. J. Manners, <u>Cereal Foods World</u>, 1985, <u>30</u>, 461.
7. D. J. Manners, <u>Carbohydr. Polym.</u>, 1987, <u>11</u>, 87.
8. S. Hizukuri and Y. Maehara, 'Biotechnology of Amylodextrinoligosaccharides' (Ed. R. B. Friedman), American Chemical Soc., Washington DC, 1989, Chapter 15, p. 212.
9. Hizukuri and Y. Maehara, <u>Carbohydr. Res.</u>, 1990, <u>206</u>, 145.
10. S. Hizukuri, S. Tabata and Z. Nikuni, <u>Starch/Stärke</u>, 1970, <u>22</u>, 338.
11. S. Tabata and S. Hizukuri, <u>Stärke</u>, 1971, <u>23</u>, 267.
12. S. Tabata, K. Nagata and S. Hizukuri, <u>Starch/Stärke</u>, 1975, <u>27</u>, 333.
13. Y. Takeda and S. Hizukuri, <u>Carbohydr. Res.</u>, 1982, <u>102</u>, 321.
14. Y. Takeda, S. Hizukuri, Y. Ozono and M. Suetake, <u>Biochim. Biophys. Acta</u>, 1983, <u>749</u>, 302.

15. J. Abe, Y. Takeda and S. Hizukuri, <u>Biochim. Biophys. Acta</u>, 1982, <u>703</u>, 26.
16. J. Abe, H. Nagano and S. Hizukuri, <u>J. Appl. Biochem.</u>, 1985, <u>7</u>, 235.
17. Y. Takeda, S. Hizukuri and B. O. Juliano, <u>Carbohydr. Res.</u>, 1986, <u>148</u>, 299.
18. S. Lansky, M. Kooi and T. J. Schoch, <u>J. Am. Chem. Soc.</u>, 1949, <u>71</u>, 802.
19. D. Borovsky, E. E. Smith, W. J. Whelan, D. French and S. Kikumoto, <u>Archiv. Biochem. Biophys.</u>, 1979, <u>198</u>, 627.
20. W. J. Whelan, <u>J. Jpn. Soc. Starch Sci.</u>, 1976, <u>23</u>, 101.

STRUCTURE-FUNCTION RELATIONSHIPS OF ß-GLUCAN HYDROLASES IN BARLEY

G.B. Fincher

DEPARTMENT OF BIOCHEMISTRY, LA TROBE UNIVERSITY, BUNDOORA, VICTORIA, 3083, AUSTRALIA

1 INTRODUCTION

Three classes of ß-glucan endohydrolases have been detected in germinating barley grain. These include the $(1\rightarrow3,1\rightarrow4)$-ß-glucan 4-glucanohydrolases (EC 3.2.1.73), the $(1\rightarrow3)$-ß-glucan glucanohydrolases (EC 3.2.1.39) and the $(1\rightarrow4)$-ß-glucan glucanohydrolases (EC 3.2.1.4). The $(1\rightarrow3,1\rightarrow4)$-ß-glucanases, which are found at high levels in the grain, hydrolyse specific $(1\rightarrow4)$-ß-glucosyl linkages in $(1\rightarrow3,1\rightarrow4)$-ß-glucans, as follows:

where G indicates the glucosyl residues, the numbers indicate $(1\rightarrow3)$- or $(1\rightarrow4)$-ß-linkages and "red" shows the reducing terminus of the polysaccharide chain. Hydrolysis of the polysaccharide is restricted to $(1\rightarrow4)$-ß-linked glucosyl residues that are themselves substituted at 03. Thus, the enzyme requires adjacent $(1\rightarrow3)$- and $(1\rightarrow4)$-ß-glucosyl residues for activity. The $(1\rightarrow3,1\rightarrow4)$-ß-glucanases exhibit an endo-hydrolytic action pattern and release $(1\rightarrow3,1\rightarrow4)$-ß-oligoglucosides of DP 3 and 4 as the major, final hydrolysis products[1]. In germinated barley, two $(1\rightarrow3,1\rightarrow4)$-ß-glucanase isoenzymes have been purified and characterized[1,2]. The isoenzymes have been designated EI and EII. Although they have identical substrate specificities and similar enzymic properties, their isoelectric points, electrophoretic mobilities and carbohydrate contents vary significantly[1,2] (Table 1).

High levels of $(1\rightarrow3)$-ß-glucan endohydrolases are also detected in extracts of germinated barley grain[3-5]. These enzymes hydrolyse $(1\rightarrow3)$-ß-glucans with an endo-hydrolytic action pattern as follows:

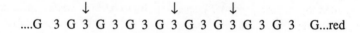

<u>Table 1</u>   Properties of barley (1→3,1→4)-ß-glucanases

Property	Isoenzyme EI	Isoenzyme EII
molecular weight	30,000	32,000
isoelectric point	8.5	10.6
pH optimum	4.7	4.7
amino acids	306	306
carbohydrate	nil	4%
N-glycosylation sites	0	1

and release laminaridextrins of DP 2 and 3 as major final products[4,5]. The enzymes can hydrolyse (1→3)-ß-glucosyl residues in mixed-linkage or substituted polysaccharides, provided regions of adjacent, unsubstituted (1→3)-linked residues are present[6]. Three distinct (1→3)-ß-glucan endohydrolase isoenzymes have been purified from extracts of germinated grain[3-5,7]. They are all basic proteins with similar electrophoretic mobilities, but exhibit subtle differences in their substrate specificities, in particular in their ability to hydrolyse substituted and branched (1→3)-ß-glucans (M. Hrmova and G.B. Fincher, unpublished data).

The third class of ß-glucan endohydrolases comprises the (1→4)-ß-glucan glucanohydrolases. These enzymes are members of the cellulase group and hydrolyse (1→4)-ß-glucosidic linkages with an endo-hydrolytic action pattern. They can also hydrolyse the (1→4) ß-linkages in barley (1→3,1→4)-ß-glucans. However, the levels of (1→4)-ß-glucanases in germinating grain are low and somewhat variable, and can be attributed, in part at least, to the presence of commensal microorganisms on the surface of the grain[8]. The cellulases are therefore unlikely to be major contributors to the degradation of endosperm (1→3,1→4)-ß-glucans, although they may play a part in the depolymerization of cellulose.

2   OCCURRENCE OF GLUCANS AND FUNCTION OF GLUCANASES

<u>Occurrence of (1→3,1→4)-ß-Glucans in Barley</u>

The (1→3,1→4)-ß-glucans are major constituents of cell walls of the starchy endosperm, where they account for up to 75% by weight of the wall[9]. The (1→3, 1→4)-ß-glucans also constitute approximately 26% by weight of aleurone walls[10,11], 20% of coleoptile walls[12], 5% of fourth internode stem cell walls[13] and 16% of walls from young leaves (N. Sakurai, personal communication). The (1→3,1→4)-ß-glucans are represented by a family of polydisperse polysaccharides which differ in their solubility, molecular size and in their fine structure[14]. The 40°C water-soluble (1→3, 1→4)-ß-glucans from barley endosperm have been subjected to detailed analyses of fine structure, molecular shape and solution properties. This polysaccharide fraction is characterized by the presence of cellotriosyl and cellotetraosyl residues linked by single (1→3)-ß-linkages[15]. These structures account for approximately 90% of the molecule[15] and it has been shown that the cellotriosyl and cellotetraosyl residues are arranged at random along the polysaccharide chain[16]. The absence of adjacent (1→3)-ß-glucosyl residues was confirmed by computer-assisted modelling studies, in which the effects of different linkage sequences on molecular shape were evaluated[17]. Whether or not other members of the cell wall (1→3,1→4)-ß-glucan family share these structural features is not clear[14].

Function of (1→3,1→4)-ß-Glucanases

In germinating barley grain the (1→3,1→4)-ß-glucanases clearly mediate in the degradation of cell walls. Starchy endosperm walls are completely removed during germination, in a process that allows α-amylases, peptidases and other hydrolytic enzymes secreted from the aleurone and/or scutellum to gain access to substrates that are initially rendered inaccessible by the cell walls[18]. The (1→3,1→4)-ß-glucanases are also likely to participate in the partial degradation of aleurone cell walls that is observed soon after the initiation of germination[19].

Table 2   Expression sites of (1→3,1→4)-ß-glucanase genes as measured by mRNA levels

Tissue	Isoenzyme EI mRNA	Isoenzyme EII mRNA
Germinated grain		
aleurone, 1 day	+	+
aleurone, 3 days	+++	++++
scutellum, 1 day	++	++
scutellum, 3 days	+	++
Vegetative tissues		
leaves, 6 days	-	-
leaves, 12 days	++	-
leaves, 20 days	++++	-
roots, 3 days	++++	-
roots, 6 days	++++	-
coleoptiles, 2 and 4 days	-	-
Effects of hormones		
gibberellic acid	enhanced (leaves)	enhanced (aleurone)
abscisic acid	no effect (leaves)	inhibited (aleurone)
auxin	enhanced (leaves)	enhanced (aleurone)

The two (1→3,1→4)-ß-glucanase isoenzymes are subject to tissue-specific differences in their expression patterns. Thus, isoenzyme EII appears to be germination-specific. It is the major (1→3,1→4)-ß-glucanase isoenzyme secreted from aleurone cells, and enzyme levels are enhanced by the addition of the phytohormone, gibberellic acid[20]. Significant levels of (1→3,1→4)-ß-glucanase are also secreted from scutellar epithelial cells, particularly in the early stages of germination[21]. However, isoenzyme EI can also be detected in the aleurone and is found in young leaves and roots[20]. In young leaves, expression of isoenzyme EI is enhanced by exogenous gibberellic acid and auxin[22]. No isoenzyme EII is found in vegetative tissues.

The presence of (1→3,1→4)-ß-glucanases in young leaves and roots suggests possible functions in aerenchyma development[23] or in the "loosening" of the wall matrix that occurs during cell expansion. The progressive, auxin-inducible decrease in (1→3,1→4)-ß-glucan content of barley coleoptile walls[12] might also be attributed to (1→3,1→4)-ß-glucanase action[18]. However, no (1→3,1→4)-ß-glucanase activity or mRNA could be detected in extracts of elongating coleoptiles[20,22] and, at this stage, neither the precise location nor the function of the isoenzyme in vegetative tissues is known. The tissue locations and influence of phytohormones on the two (1→3, 1→4)-ß-glucanase isoenzymes are summarized in Table 2.

Occurrence of (1→3)-ß-Glucans

Endogenous (1→3)-ß-glucan in the grain is restricted to small callosic deposits that are scattered through the starchy endosperm, particularly in the sub-aleurone region[24,25]. Callosic deposits are also found in other tissues during dormancy or after mechanical wounding. Although (1→3)-ß-glucans are not significant constituents of plant cell walls, they are widely distributed in the walls of fungi. Thus, (1→3)- and (1→3,1→6)-ß-glucans of diverse structure are commonly found in walls of fungal pathogens of the Basidiomycetes, Ascomycetes and Oomycetes[26].

Function of (1→3)-ß-Glucanases

The function of (1→3)-ß-glucanases in barley is not yet defined. In germinating grain levels of (1→3)-ß-glucanase activity increase markedly in the aleurone[27] and late in the germination process their specific activity can exceed that of the (1→3,1→4)-ß-glucanases (M. Hrmova and G.B. Fincher, unpublished). The apparent paucity of endogenous (1→3)-ß-glucan in the endosperm has led to the suggestion that the enzymes might be involved in the initial depolymerization and solubilization of cell wall (1→3,1→4)-ß-glucans[6]. This was given some support by reports that some (1→3, 1→4)-ß-glucan fractions contain blocks of contiguous (1→3)-linked ß-glucosyl residues that might be susceptible to hydrolysis by (1→3)-ß-glucanases[28]. The analytical procedures used in this work were subsequently questioned by Woodward *et al.*[15], who were unable to detect any adjacent (1→3)-linkages in the 40°C water-soluble (1→3,1→4)-ß-glucan. This is consistent with the inability of purified (1→3)-ß-glucanases from germinated grain and young leaves to hydrolyse this polysaccharide[4,5,7]. Thus, a role for (1→3)-ß-glucanases in cell wall degradation has yet to be convincingly demonstrated.

The abundance of (1→3)-ß-glucanases in germinating grain has led to speculation that the enzymes might function to protect the grain against microbial attack[18]. (1→3)-ß-Glucanases are widely recognized amongst the group of plant "pathogenesis-related" (PR) proteins that are synthesized in response to microbial challenge[29]. Although infection is not required for (1→3)-ß-glucanase production in the grain, the plant may synthesize the enzyme in a pre-emptive strategy in readiness for possible microbial penetration of the grain. The barley (1→3)-ß-glucanases can hydrolyse linear (1→3)-ß-glucans and substituted or branched (1→3,1→6)-ß-glucans of the type found in fungal cell walls[26]. Thus, the (1→3)-ß-glucanases could hydrolyse hyphal walls of invading fungi and cause cell lysis[30], or they could release (1→3,1→6)-ß-oligo-glucosides from the fungal walls that elicit other defense responses[31]. A variety of other enzymes and inhibitors known to be involved in plant protection have also been detected in germinated barley grain[18]. However, a role for the (1→3)-ß-glucanases in plant-pathogen interactions in barley has not been unequivocally demonstrated.

In barley there may be up to six individual (1→3)-ß-glucanase isoenzymes[32]. These individual isoenzymes are differentially expressed in various tissues and in re-sponse to fungal infection. Two isoenzymes are synthesized in the aleurone layer during germination, while other (1→3)-ß-glucanases are expressed in vegetative tis-sues, particularly in young leaves and roots (Table 3). The multiplicity of (1→3)-ß-glucanases raises questions as to why barley has evolved multiple forms of the en-zyme[32]. Possible explanations are that multiple isoforms can be independently con-trolled in different tissues, or that different isoenzymes can be targetted to different subcellular or extracellular locations. In addition, the isoenzymes have been shown to differ in their ability to hydrolyse substituted or branched (1→3)-ß-glucans (M. Hrmova and G.B. Fincher, unpublished) and a range of enzymes with subtle differences in their specificities could, in theory at least, offer protection against a wider spectrum of potential fungal pathogens[32].

Table 3 Properties of barley (1→3)-ß-glucanase isoenzyme GI-GVI

Property	GI	GII	GIII	GIV	GV	GVI
amino acids	310	306	305	327	312	315
mol. wt.	33,000	32,300	32,400	35,000	34,000	32,900
pI	8.6	9.5	9.8	10.7	7.5	4.6
tissues	roots leaves	aleurone	roots leaves	aleurone	roots leaves	?

Evolution of ß-Glucan Endohydrolases

Molecular cloning of six barley (1→3)-ß-glucanase cDNAs or genes[5,32] and the availability of genes and full-length cDNAs encoding the two (1→3)-ß-glucanase cDNAs[23,33-35] has enabled the complete amino acid sequences of the enzymes to be de-duced. Computer-assisted alignments of these sequences have revealed that the (1→3)-

ß-glucanases and (1→3,1→4)-ß-glucanases show approximately 50% positional identity overall, and that much higher values are observed in localized regions of the two classes of enzymes, particularly towards their COOH-termini[32]. These similarities, together with striking similarities in their gene structures, has led to suggestions that the (1→3)-ß-glucanases and (1→3,1→4)-ß-glucanases are representatives of a single "super-gene" family in which members share a common evolutionary history[32].

In view of the similar gene structures and the related substrate specificities of the (1→3)- and the (1→3,1→4)-ß-glucanases, it is tempting to speculate that the two classes of enzymes evolved from a single ancestral enzyme that was capable of hydrolysing both polysaccharides, as is observed for the ß-glucan endohydrolase (EC 3.2.1.6) from *Rhizopus arrhizus*. The subsequent appearance of enzymes with distinct substrate specificities might also reflect co-evolution with cell wall polysaccharides of plants and fungi, and the specialization in function of the two classes of enzymes. Thus, the (1→3)-ß-glucanases retain a role in the metabolism of plant cell walls, while the (1→3)-ß-glucanases might offer a protective role to the plant through their ability to hydrolyse fungal cell walls.

2 ACKNOWLEDGEMENTS

I am indebted to Danny Doan, Peter Høj, Maria Hrmova, Nada Slakeski and Peilin Xu for their enthusiasm in characterizing the enzymes and genes described here. The support of the Australian Research Council and the Grains Research and Development Corporation is gratefully acknowledged.

3 REFERENCES

1. J.R. Woodward and G.B. Fincher, Carbohydr. Res., 1982, 106, 111.
2. J.R. Woodward and G.B. Fincher, Eur. J. Biochem., 1982, 121, 663.
3. G.M. Ballance and I. Svendsen, Carlsberg Res. Commun., 1988, 53, 411.
4. P.B. Høj, A.M. Slade, R.E.H. Wettenhall and G.B. Fincher, FEBS Lett., 1988, 230, 67.
5. P.B. Høj, D.J. Hartman, N.A. Morrice, D.N.P. Doan and G.B. Fincher, Plant Molec. Biol., 1989, 13, 31.
6. G.M. Ballance and D.J. Manners, Phytochemistry, 1978, 17, 1539.
7. J. Wang, P. Xu and G.B. Fincher, Eur. J. Biochem., 1992, 209, 103.
8. J.L. Hoy, B.J. Macauley and G.B. Fincher, J. Inst. Brew., 1981, 87, 77.
9. G.B. Fincher, J. Inst. Brew., 1975, 81, 116.
10. A. Bacic and B.A. Stone, Aust. J. Plant Physiol., 1981, 8, 453.
11. A. Bacic and B.A. Stone, Aust. J. Plant Physiol., 1981, 8, 475.
12. N. Sakurai and Y. Masuda, Plant Cell Physiol., 1978, 19, 1217.
13. A. Kokubo, S. Kuraishi and N. Sakurai, Plant Physiol., 1989, 91, 876.
14. J.R. Woodward, D.R. Phillips and G.B. Fincher, Carbohydr. Polymer., 1988, 8, 85.
15. J.R. Woodward, G.B. Fincher and B.A. Stone, Carbohydr. Polym., 1983, 3, 207.
16. R.G. Staudte, J.R. Woodward, G.B. Fincher and B.A. Stone, Carbohydr. Polym., 1983, 3, 299.

17. G.S. Buliga. D.A. Brant and G.B. Fincher, Carbohydr. Res., 1986, 157, 139.
18. G.B. Fincher, 'Barley: Genetics, molecular biology and biotechnology', (Ed. P.R. Shewry), CAB International, Wallingfond, 1992, pp 413.
19. F. Gubler, A.E. Ashford and J.V. Jacobsen, Planta, 1987, 172, 155.
20. N. Slakeski and G.B. Fincher, Plant Physiol., 1992, 99, 1226.
21. G.I. McFadden, B. Ahluwalia, A.E. Clarke and G.B. Fincher, Planta, 1988, 173, 500.
22. N. Slakeski and G.B. Fincher, FEBS Lett., 1992, 306, 98.
23. N. Slakeski, D.C. Baulcombe, K.M. Devos, B. Ahluwalia, D.N.P. Doan and G.B. Fincher, Molec. Gen. Genetics, 1990, 224, 437.
24. R.G. Fulcher, G. Setterfield, M.E. McCully and P.J. Wood, Aust J. Plant Physiol., 1977, 4, 917.
25. A.W. MacGregor, G.M. Ballance and L. Dushnicky, Food Microstructure, 1989, 8, 235.
26. J.G.H. Wessels and J.H. Sietsma, 'Encyclopedia of Plant Physiol. Plant Carbo-hydrates II. (NS)', (Eds. W. Tanner and F.A. Loewus), Springer-Verlag, Berlin, 1981, Vol. 13B, pp. 352.
27. G.M. Ballance, W.O.S. Meredith and D.E. Laberge, Can. J. Plant Sci., 1976, 56, 459.
28. O. Igarashi and Y. Sakurai, Agric. Biol. Chem., 1966, 30, 642.
29. T. Boller, 'Plant-Microbe Interactions: Molecular and Genetic Perspectives', (Eds. T. Kosuge and E.W. Nester), Macmillan, New York, 1987, Vol. 2, pp. 385.
30. R. Leah, H. Tommerup, I. Svendsen and J. Mundy, J. Biol. Chem., 1991, 266, 1564.
31. R.A. Dixon and C.J. Lamb, Ann. Rev. Plant Physiol. Plant Mol. Biol., 1990, 41, 339.
32. P. Xu, J.Wang and G.B. Fincher, Gene, 1992, 120, 157.
33. G.B. Fincher, P.A. Lock, M.M. Morgan, K. Lingelbach, R.E.H. Wettenhall, A. Brandt and K-K. Thomsen, Proc. Nat. Acad. Sci. (USA), 1986 83, 2081.
34. J.C. Litts, C.R. Simmons, E.E. Karrer, N. Huang and R.L. Rodriguez, Eur. J. Biochem., 1990, 194, 831.
35. N. Wolf, Plant Physiol., 1991, 96, 1382.

STARCH SYNTHESIS IN TRANSGENIC PLANTS

L. Willmitzer, J. Koßmann, B. Müller-Röber and U. Sonnewald

INSTITUT FÜR GENBIOLOGISCHE FORSCHUNG BERLIN GMBH,
IHNESTRASSE 63, W-1000 BERLIN 33, GERMANY

1 INTRODUCTION

Plant biotechnology has bloomed within the last decade. Major breakthroughs are the development of methods allowing the transfer of isolated and well-defined genes into plant cells, the regeneration of intact and fertile transgenic plants, the identification of regulatory structures within plant genes controlling the expression of these genes on the transcriptional level as well as the identification of structures in proteins directing them to certain subcellular organelles. In addition the possibility of inhibiting the expression of endogenous genes via introducing genes coding an anti-sense RNA has become a routine method.

It is obvious that this methodology can not only be used for a better understanding of the plant but also to improve plants in traits wanted by mankind. Thus massive efforts are presently undertaken aiming at modifying plants with respect to their biosynthetic capacity trying to produce new or modified compounds in higher plants. This idea of using plants as a bioreactor is not only a challenging scientific question but also might create a new market for the farmer which at least in Western Europe and North America is politically desirable.

In the first part of this contribution we will briefly review the recently developed methodology allowing transgenic plants to be created. In the second part we will describe first applications of this technique with respect to modifying starch quality and quantity in transgenic plants.

2 TECHNIQUES DEVELOPED FOR THE CREATION OF TRANSGENIC PLANTS CHANGED IN A CERTAIN PARAMETER

Transfer of DNA into Plant Cells and Regeneration of Intact and Fertile Plants

Transfer of isolated DNA into plant cells is now a routine procedure which works for many different plants including crop species. Several methods exist to transfer DNA into cells. The ability of the gram-negative soil bacterium Agrobacterium tumefaciens to introduce genetic material into the nucleus of at least most of the dicotyledonous plants (cf. 1 for a review) is still the preferred method in the case of transforming dicotyledonous plants. Second to this is the direct transformation of cell-wall less protoplasts of plants by naked DNA present in the medium in the presence of either polyethylene glycol or calcium (cf. 2 for a review). Numerous other possibilities have been described for the delivery of DNA into cells including microinjection[3], liposome fusion[4] or the biolistics approach[5]. Whereas it is therefore no major problem to obtain single cells or undifferentiated cell clumps genetically being altered due to gene transfer, it turns out to be a much more difficult problem to obtain intact and fertile transgenic plants.

Thus it is probably fair to say that at the present stage the number of plant species and genotypes accessible to transformation is no longer limited by the methodology to transfer single isolated genes into plant cells but rather by the ability/inability to re-generate intact and fertile plants out of genetically transformed plant cells. The different strategies to overcome this problem go along with using different vector systems/delivery systems for transferring DNA into plant cells and are connected to the plant species to be transformed resp. its tissue culture ability.

Controlled Expression of Foreign Genes in Transgenic Plants

One important aspect of plant gene technology and the production of transgenic plants is the controlled and predetermined expression of foreign genes. A special desire might e.g. be to produce a certain enzyme/protein only during certain developmental phases of a plant such as embryogenesis or germination, in certain organs e.g. seeds, leaves, tubers, roots or flowers, in certain tissues e.g. epidermis, phloem, tapetum, mesophyll cells, meristematic tissues or only during certain environmental conditions e.g. in the light, in the dark, upon pathogen attack, upon wounding, under drought conditions etc.

Due to the fact that the identification and characterization of plant genes has been one of the major research areas in Plant Molecular Biology during the last couple of years, a clear concept has developed allowing the construction of chimeric genes expressing the coding sequence in a predetermined manner.

Generally it has been observed that sequences located 5´ to the RNA-coding sequence of the respective gene contain many if not all of the signals necessary to ensure the correct expression. This has been proven by taking the 5´-upstream region of plant genes, fusing it to a marker gene encoding an easily scorable product and, after addition of a poly-adenylation signal, transfer the chimeric gene back to the plant using either one of the above mentioned techniques. A list of typical representatives of various types of promoters identified in different plants is given[6].

Despite the fact that for many promoters it is proven that they retain their specificity of expression in heterologous plants, the level of expression of one and the same gene present in independent transformants as a rule varies widely[7]. This phenomenon, which is not specific for plants but has been observed in other transgenic eucaryotes too, is probably due to sequences neighbouring the transferred gene after its integration. Surprisingly, although these neighbouring sequences strongly influence the quantitative level of expression of the transferred gene, as a rule, they do not at all influence the qualitative expression pattern.

Targetting of Foreign Proteins into Certain Subcellular Compartments

A characteristic feature of a eucaryotic cell is the presence of various compartments which e.g. serve to separate anabolic from catabolic metabolic pathways. In order to be able to change plants in a predetermined way one may wish to not only control the formation of a protein/enzyme with respect to a certain tissue, developmental or environmental condition (as guaranteed by the promoter used) but also to control its subcellular location within the cell. Thus the idea of targetting proteins to a predetermined destination within the cell is a very attractive one.

Trafficking of nuclear-encoded proteins within a eucaryotic cell occurs in several ways depending upon the final destination of the protein. In the case of proteins destined to the mitochondria or the chloroplast they are synthesized on free polysomes and given an N-terminal extension (transit peptide) which directs them into either the chloroplast or the mitochondrion where the transitpeptide is cleaved off during the import process. A large bulk of the proteins is destined to move through the endoplasmic reticulum/Golgi pathway. They either become secreted into the extracellular space, become part of the plasma membrane, stay in the ER/Golgi complex or move into the vacuole. This large and complex class of proteins is believed to be cotranslationally imported into the lumen of the endoplasmic reticulum as a first step with further sorting to occur in the cis- and trans-Golgi. Again all these proteins contain a N-terminal extension (signal peptide) necessary to direct the nascent polypeptide chain to specific receptors present on the membrane of the endoplasmic reticulum.

By fusing DNA-sequences encoding the N-terminal extension of proteins destined to the chloroplast or the mitochondrion to reporter genes, chimeric genes have been constructed which, after addition of plant promoters and poly-adenylation signals resulted in the formation of fusion proteins in transgenic plants. Moreover these proteins were observed not to stay within the cytosol but rather to move into different organelles, the type of organelle being determined by the origin of the N-terminal sequence added[8,9]. As more and more transit peptides from nuclear encoded, chloroplast or mitochondrion located proteins have been identified, numerous possibilities for the creation of chimeric proteins exist which makes the targetting of foreign proteins into either chloroplasts or mitochondria of transgenic plants a fairly straightforward experiment.

The situation is slightly different for proteins cotranslationally taken up into the endoplasmic reticulum and subsequently distributed among the vacuole, the plasma membrane, the extracellular space and the ER-Golgi complex. With respect to directing proteins into the lumen of the endoplasmic reticulum, the same scheme could be applied. Fusion of the N-terminal extension (signal peptide) of vacuolar or secreted proteins to the foreign protein resulted in its uptake into the ER[10]. With respect to the further sorting however a couple of questions still have to be answered. Thus there is only one example where in transgenic plants a foreign protein could be directed into the vacuoles[11] by fusing a very large part (146 aa) of a vacuolar plant protein to the N-terminus of the reporter protein. In other cases it has been shown that the last 10 - 12 amino acids located at the C-terminus would be responsible for vacuolar targetting (cf. 12 for a review).

Secretion of proteins into the extracellular space seems to occur if no further signal in addition to the signal peptide is present. Secretion of chimeric proteins only given the signal peptide of a plant protein has been demonstrated in transgenic plants[13,14].

Inactivation of the Expression of Endogenous Genes in Transgenic Plants

In addition to being able to add new traits to a plant, the ability to inhibit the manifestation of some unwanted characteristics obviously represents a very desirable trait. This has been achieved in an efficient way via introduction of a chimeric gene giving rise to the formation of an RNA strand complementary to the strand produced from the endogenous gene in the transgenic plant (anti-sense RNA approach)[15]. The anti-sense RNA is complementary to the target mRNA which will be inhibited with respect to its formation/biological activity. It is assumed (though not proven) that a duplex RNA will be formed between the mRNA resulting from the endogenous gene and the anti-sense RNA being produced from the introduced gene which, by an unknown mechanism, leads to the inactivation and the disappearance of both the sense- and the anti-sense mRNA.

The anti-sense RNA mediated inhibition of the expression of endogenous genes has been demonstrated for an increasing number of genes[16].

One further advantage of the anti-sense RNA inhibited gene expression is that, as a rule, plants independently transformed with the anti-sense RNA coding gene give rise to different degrees of inhibition. Thus it is not only possible to suppress the expression of the endogenous gene to near completeness (as is the case for gene disruption via homologous recombination) but also to obtain transgenic plants differing widely in their degree of inhibition.

3   MODULATION OF STARCH QUALITY AND QUANTITY IN TRANS-GENIC PLANTS

Molecular Approaches to Reduce Starch Synthesis in Potato Tubers

ADP-glucose formed by the enzyme ADP-glucose pyrophosphorylase from glucose-1-phosphate and ATP is the precursor used by the starch synthases for the formation of linear α-1,4-glucans. ADP-glucose pyrophosphorylase has been described as the enzyme controlling the rate of starch biosynthesis[17]. We therefore decided to try to inhibit starch biosynthesis via inhibiting the activity of the ADP-glucose pyrophosphorylase. To this end cDNA's encoding both subunits of the ADP-glucose pyrophosphorylase were cloned from potato using heterologous probes from maize[18].

Subsequently, chimeric genes expressing an anti-sense RNA from both cDNA's were introduced into potato by Agrobacterium-mediated gene transfer and transgenic potato plants regenerated.

These plants were subsequently analyzed on the RNA, protein and enzyme activity level for inhibition of the ADP-glucose pyrophosphorylase gene expression. Out of a collection of 50 independently transformed potato plants, 4 transgenic plants were identified which displayed only 2-5% of the wild-type activity of the ADP-glucose pyrophosphorylase. In parallel to this reduction in the enzymic activity, the tuber starch content was reduced to 2-5% of wild-type levels[19].

As a result of the inhibition of starch biosynthesis, a dramatic change in the phenotype of the tubers formed was observed. Whereas in the case of wild-type plants the average number of tubers grown under standardized greenhouse conditions amounts to about 8-15, potato plants unable to synthesize starch produced 70-90 tubers. A second significant change concerns the size of the tubers, which is significantly reduced in the starch-less tubers compared to tubers from wild-type plants.

Inhibition of starch biosynthesis in tubers of transgenic potato plants thus has profound effects on number and size of this sink organ. The increase in number and the decrease in size of an individual tuber can be interpreted as indicating a reduction in sink strength. This interpretation is based on the observation that upon induction of the tuberization process in potato, multiple tuber-forming primordia are initiated. However in wild-type plants only a minor proportion of these tuber primordia develop into mature tubers[20]. In wild-type plants normally only one or two tubers are formed per stolon; the anti-sense plants developed up to ten. In addition, alternating tuber-stolon structures were observed in plants which had low ADP-glucose pyrophosphorylase activity, a phenomenon which very rarely occurs in control plants grown under standard greenhouse conditions.

Accepting the concept of competing sinks, this can be explained by assuming that very early during tuber growth some tubers have gained a competitive advantage and can thus outcompete the other tuber primordia with respect to carbohydrate which therefore will stop their further development. The observation that starch-less potato plants form significantly more tubers indicates that the creation of a competitive advantage with respect to sink strength could well be due to the action of the ADP-glucose pyrophosphorylase, making this enzymic activity therefore a possible determinator for sink strength. It is obvious that this interpretation is still speculative at this level and that further experiments are needed to further substantiate this model.

Molecular Approaches to Increase Starch Quantity in Transgenic Tubers of Potato

A major breakthrough with respect to quantity was recently reported by a group working at Monsanto. The enzyme ADP-glucose pyrophosphorylase represents one of the major control points in the starch biosynthesis pathway. The plant enzyme is subject to feed-back inhibition by several metabolites including phosphate[17]. Glycogen-producing bacteria contain a very similiar enzyme also catalyzing the formation of ADP-glucose from ATP and glucose-1 phosphate. By expressing a mutated form of the E. coli ADP glucose pyrophosphorylase which is changed in various of its allosteric properties in amyloplasts of transgenic potato plants, Kishore et al[21] were able to create potato plants which in their tubers display increased starch and a higher dry matter content. On average these tubers contain 24% higher level of dry matter relative to controls with some lines containing nearly 50% higher levels.

Molecular Approaches to Alter the Starch Quality in Potato Tubers - Anti-Sense Inhibition of the Granule Bound Starch Synthase

Starch is synthesized by the combined action of the starch synthases and the branching enzymes. The starch synthases catalyze the polymerisation of ADP-glucose to linear α-1,4-glucans, whereas the branching enzymes introduce branch points by releasing small segments from a linear chain and reattaching them to a glucan through the formation of α-1,6-glucosidic bonds, which leads to the formation of amylopectin.

Mutants lacking one isoform of the starch synthases, the insoluble or granule bound form, synthesize an essentially amylose free-starch[17].

A cDNA clone encoding the GBSS from potato (a kind gift from M. Hergersberg and U. Wienand, Cologne) was expressed in anti-sense orientation in transgenic potato plants driven by a tuber-specific promotor. Plants which showed almost complete reduction of the GBSS protein synthesized, like the classical mutants, starch which has a very low amylose content (less than 5%). This is in agreement with data from another group persuing the same approach[22]. The total starch content in the tubers was not significantly altered. Field trials showed that the anti-sense RNA mediated inhibition of the amylose formation was stable[23].

Anti-Sense Inhibition of the Starch Branching Enzyme

Branching enzyme (BE) normally appears in two isoforms in plants. Mutants of maize and pea have been characterized where the activity of one isoform of the enzyme is missing, leading to an increase of the amylose content (from 20-30% up to 70%) of the synthesized starch[17]. In potato tubers only one form of BE has been described[24].

Potato plants were transformed with a construct carrying a cDNA for BE[25] in anti-sense-orientation behind the 35S CaMV promotor. Those plants which did not express detectable amounts of BE, as determined by Western blotting, were selected for further analysis. Surprisingly neither the amylose content of the starch in the tubers of these plants, nor the total starch content of the tubers was altered.

One possible explanation is that there is another isoform of the enzyme in potato tubers, which has not been characterized so far. Evidence for that is that the residual branching enzyme activity in the mutant plants lies between 10-20% of control plants, which is significantly higher than the BE protein levels which can be detected on western-blots.

4 CONCLUSIONS

As briefly described in this contribution, the tools have been developed to modify higher plants by genetic engineering methods which among other aspects will also allow changes in starch quality and quantity in transgenic plants. Whether or not this will lead to a new technology of economic importance will mainly depend upon the public acceptance of transgenic plants.

5 REFERENCES

1. P. Zambryski, J. Tempe and J. Schell, Cell, 1989, 56, 193.
2. I. Potrykus, Ann.Rev. Plant Physiol. Plant Mol. Biol., 1991, 42, 205.
3. A. Crossway, J.V. Oakes, J.M. Irvine, B. Ward, V.C. Knauf and C.K. Shewmaker, Mol. Gen. Genetics, 1986 202, 179.
4. M. Caboche, Physiol. Plant 1990 79, 173.
5. T.M. Klein, E.D. Wolf, R.D. Wu and J.C. Stanford, Nature, 1987, 327, 70.
6. J. Edwards and G. Coruzzi, Ann. Rev. Genetics, 1990, 24, 275.
7. L. Willmitzer, Trends in Genetics, 1988, 4, 13.

8. K. Keegstra, L. Olsen and S. Theg, <u>Ann. Rev. Plant Physiol.Mol. Biol.</u>, 1989, <u>40</u>, 471.
9. M. Boutry, F. Nagy, C. Poulson, K. Aoyago and N. Chua, <u>Nature</u>, 1987, <u>328</u>, 340.
10. G. Iturriaga, R. Jefferson and M. Bevan, <u>The Plant Cell</u>, 1989, <u>1</u>, 381.
11. U. Sonnewald, M. Brauer, A. von Schaewen, M. Stitt and L. Willmitzer, <u>The Plant Journal</u>, 1991, <u>1</u>, 95.
12. F. Sebastiani, L. Farrell, M. Vasquez and R. Beachy, <u>Eur. J. Biochem.</u>, 1991, <u>199</u>, 441.
13. C. Dorel, T. Voelker, E. Herman and M. Crispeels, <u>J. Cell Biol.</u>, 1989, <u>108</u>, 327.
14. A. von Schaewen, M. Stitt, R. Schmidt, U. Sonnewald and L. Willmitzer, <u>EMBO J</u>, 1990, <u>9</u>, 3033.
15. A. Moffat, <u>Science</u>, 1991, <u>253</u>, 510.
16. A. van der Krol, J. Mol and A. Stuitje, <u>BioTechniques</u>, 1988, <u>6</u>, 958.
17. J. Preiss, 'The biochemistry of plants', Academic Press, New York, 1988,Vol. 14, pp.181.
18. B. Müller-Röber, J. Koßmann, C. Hannah, L. Willmitzer and U. Sonnewald, <u>Mol. Gen. Genetics</u>, 1990, <u>224</u>, 136.
19. B. Müller-Röber, U. Sonnewald and L. Willmitzer, <u>EMBO J.</u>, 1992, <u>11</u>, 1229.
20. P.H. Li, 'Potato Physiology', Academic Press, New York, 1985.
21. S. Kishore, International Patent Application WO 91/19806, 1991.
22. R. Visser, I. Somhorst, G. Kuipers, N. Ruys, W. Feenstra and E. Jacobsen, <u>Mol. Gen. Genetics</u>, 1991, <u>225</u>, 289.
23. G.J. Kuipers, J. Vreem, H. Meyer, E. Jacobsen, W. Feenstra and R. Visser, <u>Euphytica</u>, 1992, <u>59</u>, 83.
24. D. Borovsky, E.E. Smith and W.J. Whelan, <u>Eur. J.Biochem.</u>, 1975, <u>59</u>, 615.
25. J. Koßmann, R.G.F. Visser, B. Müller-Röber, L. Willmitzer and U. Sonnewald, <u>Mol.Gen.Genet.</u>, 1991, <u>230</u>, 39.

Rheology

CONSTITUTIVE MODELS FOR DILUTE AND CONCENTRATED FOOD BIOPOLYMER SYSTEMS

J. L. Kokini

DEPARTMENT OF FOOD SCIENCE, RUTGERS, THE STATE UNIVERSITY OF NEW JERSEY, P.O. BOX 231, NEW BRUNSWICK, NEW JERSEY 08903-0231, USA

1 INTRODUCTION

As early as Einstein[1] researchers have attempted to relate rheological measurements to the structural unit responsible for the observed behaviour. In order to understand the relationship between rheological properties and their structure idealizations of their conformation is necessary. A typical example is the freely jointed chain consisting of springs and beads[2]. Such idealizations are the domain of molecular rheology. These idealizations lead to models which describe stresses developed in materials as a result of an applied deformation history. These models are usually able to describe the relationship between all components of stress and strain as well as strain rate and they are referred to as constitutive models[3-5].

The use of constitutive models in reference to food materials is not new. Linear viscoelastic models such as the Maxwell or Voigt models as well as the generalized Maxwell model and the four parameter models have been used for materials such as wheat flour doughs[6], ice cream, mayonnaise[7], salad dressing emulsions[8-10]. However, these models are only applicable for very small deformations and often are limited in scope.

When the Maxwell model or generalized Maxwell model are redefined in convected coordinates a family on non-linear models result which have greater usefulness because they are able to describe more realistic flows[2]. Examples are the Oldroyd, Spriggs, Carreau as well as the Bogue and Chen models. Bagley[4] found the upper convected Maxwell model adequate to explain the biaxial extensional flow of cheese as well as wheat flour doughs. Bagley and Christiansen[11] have also used the K-BKZ constitutive model to characterize stress relaxation behaviour after uniaxial compression of doughs.

Kokini et al.[12] used the Bird-Carreau constitutive model to predict steady shear and small amplitude oscillatory properties of guar and carrageenan gum solutions. Kokini and Plutchok[13] extended this approach to gum blends of carboxymethylcellulose (CMC) and guar gum. Dus and Kokini[14] used the same model to predict steady shear and small amplitude oscillatory properties of 40% moisture hard wheat flour doughs.

Dilute solution molecular theories are also beginning to find applications in food polymer rheology. Chou et al.[15] tested the Rouse[16], Zimm[17] theories for random coils and the Marvin and McKinney[18] theory for rod-like molecules in reference to citrus pectin solutions. In the Rouse concept the molecules are totally free draining with no hydrodynamic interactions. In the Zimm approximation on the other hand, it is assumed that hydrodynamic interactions have a significant effect and are taken into account when calculating the spectrum of relaxation times. At the other extreme it is possible to develop molecular models which approximate the flow behaviour of elongated molecules as rigid rods. Examples include Yamakawa's cylinder, the shishkebab model and Ulmann's cylinder etc.[19] Marvin and McKinney also developed a model approximating rod-like behaviour. When Kokini et al.[20] compared the dilute solution behaviour of apple pectins with the Zimm, Rouse and rod-like theories the Zimm model gave the best approximation.

The objective of this paper is to review some of the molecular theories mentioned above and to present methods and data which is applicable to dilute and concentrated food biopolymer systems.

Dilute Solution Molecular Theories

The equations to predict the reduced moduli and relaxation time of flexible random coil molecules of Rouse and Zimm type are as follows:

$$[G']_R = \sum_{p=1}^{n} \frac{\omega^2 \tau_p^2}{(1+\omega^2 \tau_p^2)}$$

and

$$[G'']_R = \sum_{p=1}^{n} \frac{\omega \tau_p}{1+\omega^2 \tau_p^2}$$

where $[G']_R$ is reduced storage modulus given by:

$$\left[G'\right]_{\substack{R \\ c \to 0}} = \frac{G'M}{CRT}$$

and G' is the storage modulus of the dilute solution, M is molecular weight, C is concentration, R is the ideal gas constant and T is the absolute temperature.

The reduced loss modulus $[G'']_R$ is then given by:

$$\left[G''\right]_{\substack{R \\ c \to 0}} = \frac{(G''-\omega \eta_s)M}{CRT}$$

where η_s is the solvent viscosity, ω is the frequency of the oscillation and G'' is the loss modulus.

The relaxation time necessary to calculate the reduced storage and loss moduli is given as follows:

$$\tau_p = \frac{K_p [\eta] \eta_s M}{CRT}$$

To predict rigid rod behaviour several theories are available. The general form of these predictions are as follows:

$$[G']_R = \frac{m_1 \omega^2 \tau^2}{1 + \omega^2 \tau^2}$$

$$[G'']_{\substack{R \\ c \to 0}} = \omega \tau \left[\frac{m_1}{(1 + \omega^2 \tau^2)} + m_2 \right]$$

and

$$\tau = \frac{m [\eta] \eta_s M}{RT}$$

and m is given by

$$m = (m_1 + m_2)^{-1}$$

Table 1 shows the values of these constants for five different rigid rod models. The predicted reduced moduli from the theory of Marvin and McKinney for rigid dumbbells[18] as a function of $\omega\tau$ are given in Figure 1 and the predicted moduli for the random coil theories of Rouse and Zimm are shown in Figure 2. At high frequencies the reduced moduli of the Rouse theory become equal and increase together with a slope of 1/2 while those in the Zimm theory remain unequal and increase in a parallel manner with a slope of 2/3.

Table 1   Empirical constants m_1 and m_2 for the elongated rigid rod model

Model	m_1	m_2	m
Cylinder (Yamakawa, 1975)	0.60	0.29	1.15
Cylinder (Ullman, 1969)	0.46	0.16	1.61
Rigid Dumbell (Marvin and McKinney, 1965)	0.60	0.40	1.00
Prolate Ellipsoid (Cerf, 1952)	0.60	0.24	1.19
Shishkebab (Kirkwood and Auer, 1951)	0.60	0.20	1.05

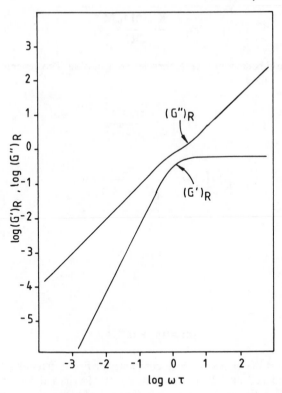

<u>Figure 1</u>   Reduced moduli predictions for the rigid rod theory of Marvin and McKinney[18]

Concentrated Solution/Melt Theories

The Bird-Carreau Model.   The Bird-Carreau model is an integral model which involves taking an integral over the entire deformation history of the material[21]. James[22] was first to develop a mathematical model for statistical properties of a molecular network[23] which consists of polymer chains physically crosslinked, forming macromolecular structure. The Bird-Carreau integral constitutive model is based on the Carreau constitutive theory of molecular networks which is able to explain viscoelastic behaviour by assuming deformation creates and destroys temporary crosslinks[24].

The Bird-Carreau model employs the use of four empirical constants and the zero shear limiting viscosity of the solutions. Two constants, α_1 and λ_1 are obtained from a logarithmic plot of η vs. $\dot{\gamma}$ and the other two constants α_1 and λ_2 are obtained from a logarithmic plot of η' vs. ω as shown in Figure 3.

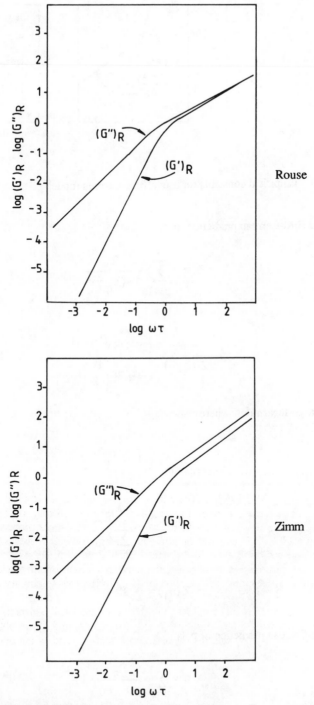

<u>Figure 2</u>   Reduced moduli predictions for flexible random coils as proposed by Rouse[16] and Zimm[17]

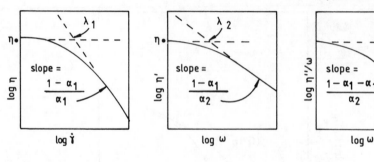

<u>Figure 3</u>   Empirical constants for use in Bird-Carreau model

The Bird-Carreau prediction for η is:

$$\eta = \sum_{p=1}^{\infty} \frac{\eta_p}{1+(\lambda_{1p}\dot{\gamma})^2}$$

$$\eta = \frac{\pi\,\eta_0}{Z(\alpha_1)-1} \cdot \frac{(2^{\alpha 1}\lambda_1\dot{\gamma})^{(1-\alpha_1)/\alpha_1}}{2\alpha_1 \sin\left[\frac{1-a_1}{2a_1}\cdot\pi\right]}$$

at large shear rates, where

$$\lambda_{1p} = \lambda_1\left[\frac{2}{p+1}\right]^{\alpha_1}$$

$$\eta_p = \eta_0 \; \frac{\lambda_{1p}}{\sum_{p=1}^{\infty}\lambda_{1p}}$$

$$Z(\alpha_1) = \sum_{k=1}^{\infty} K^{-\alpha_1}$$

The Bird-Carreau prediction of η' is:

$$\eta' = \sum_{p=1}^{\infty} \frac{\eta_p}{1+(\lambda_{2p}\omega)^2}$$

and,

$$\eta' = \frac{\pi\ \eta_o}{Z(\alpha_1)-1} \cdot \frac{(2\alpha_2\ \lambda_2\omega)^{(1-\alpha_1)/\alpha_2}}{2\alpha_2\ \sin\left[\frac{1+2\alpha_2-\alpha_1}{2\alpha_2} \cdot \pi\right]}$$

at high frequencies.

The Bird-Carreau Model for η''/ω is:

$$\eta''/\omega = \sum_{p=1}^{\infty} \frac{\eta_p\lambda_{2p}}{1+(\lambda_{2p}\omega)^2}$$

and,

$$\eta''/\omega = \frac{2^{\alpha_2}\lambda_2\pi\eta_o}{Z(\alpha_1)-1} \cdot \frac{(2^{\alpha_2}\lambda_2\omega)^{(1-\alpha_1-\alpha_2)/\alpha_2}}{2\alpha_2\ \sin\left[\frac{1+\alpha_2-\alpha_1}{2\alpha_2} \cdot \pi\right]}$$

at high frequency, where,

$$\lambda_{2p} = \lambda_2\left[\frac{2}{p+1}\right]^{\alpha_2}$$

Doi and Edwards[25] viscoelasticity is explained by considering entanglements within the polymer network. In this theory a model chain (or primitive path) is constructed which describes molecular motions in a densely populated system using appropriate assumptions. It is first assumed that each polymer chain moves independently in the mean field imposed by the other chains. The mean field is represented by a three dimensional cage. In this cage each polymer is confined to a tube like region surrounding it as shown in Figure 4. The primitive chain can move randomly forward or backward only along itself.

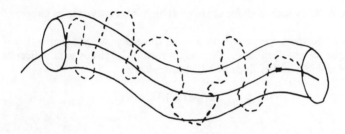

Figure 4   The cage and tube model

To define dynamic properties under flow a slip-link network model is introduced. The junctions of slip-links are assumed to be not permanent cross-links but small rings through which the chain can pass freely. This model gives a simple molecular mechanism of the breakage and creation process as it occurs by the sliding motion of the chain through the slip links as shown in Figure 5. In a system of highly entangled polymers the molecular motion of a single chain can be divided into two types:

i) the small scale wiggling motion which does not alter the topology of the entanglement; and

ii) the large scale diffusive motions which changes the topology.

Feq

<u>Figure 5</u>   The slip-link model

The time scale of the first motion is essentially the Rouse relaxation time[26]. The time scale of the second motion denoted by Td is a renewal time of the topology of a single chain and is proportional to M^3 [27]. The Doi-Edwards theory is only concerned with the motion of the second type. For a polydisperse system the theory has been modified by Rahalkar et al.[28] The following results are relevant to the storage and loss moduli (G' and G") of a monodisperse polymer.

$$G'(\omega) = \frac{8}{\pi^2} G^0{}_N \sum_{p=1,\,odd}^{\infty} \left[(\omega T_1)^2/p^6\right]/\left[1+(\omega T_1)^2/p^4\right]$$

$$G''(\omega) = \frac{8}{\pi^2} G^0{}_N \sum_{p=1,\,odd}^{\infty} \left[(\omega T_1)/p^6\right]/\left[1+(\omega T_1)^2/p^4\right]$$

where $G^0{}_N$ is the plateau modulus obtained at high frequency, p is an integer.

For a polydisperse polymer with a given molecular weight distribution if f (μ) is the molecular weight distribution, the weight fraction of chains with molecular weight between M and M + dM is given by W(M)dM, where

$$W(M) = 1/M_n \, f(\mu),$$

where

μ is the dimensionless molecular weight ($=M/M_n$).

For this case the storage and loss modulus are given by:

$$G'(\omega) = G^0_N \int_0^\infty 8/\pi^2 \sum_{P+1,odd}^\infty \left[(\omega T_1)\mu^6 f(\mu)/p^6\right] / \left[1+(\omega T_1)^2 \mu^6/p^4\right] d\mu$$

and

$$G'(\omega) = G^0_N \int_0^\infty 8/\pi^2 \sum_{P+1, odd}^\infty \left[(\omega T_1)\mu^3 f(\mu)/p^4\right] / \left[(1+\omega T_1)^2 \mu^6/p^4\right] d\mu$$

where G^0_N, the plateau modulus is given by

$$G^0_N = G_0 ave/5$$

Applications of Constitutive Models to Food Systems

Dilute Solution Theories.   We used the theories of Rouse and Zimm as well as rodlike theories to study the conformation of apple, tomato and citrus pectins in solution. Estimation of intrinsic moduli [G'] and [G"] necessitates measurement of the storage modulus G' and the loss modulus G" at several concentrations in the dilute solution region and then extrapolation to zero concentration. Small amplitude oscillatory measurements were conducted in 0.25 M NaCl to achieve maximum dissociation between chains in 95% glycerol and 5% deionized water mixtures as the solvent. The first key test is the success of the extrapolation used to obtain intrinsic moduli. The calculated intrinsic loss and storage moduli are shown in Table 2 for an apple pectin (D.E 73.5%). It can be concluded from the table that the extrapolation procedure is quite reliable.

Table 2   Extrapolation of dynamic data to obtain intrinsic moduli

NSS*	NSS		NSS	
log $\omega\tau$	(G')	R (2)	(G")	R (2)
-0.92	0.19	0.67	0.49	0.96
-0.72	0.26	0.86	0.96	0.98
-0.52	0.28	0.94	-7.90	0.98
-0.32	0.52	0.92	1.83	0.98
-0.12	0.62	0.94	2.25	0.98
0.08	0.93	0.96	3.33	0.98
0.28	1.39	0.98	4.92	0.98
0.48	2.15	0.98	7.18	0.98
0.68	3.29	0.98	10.67	0.98
0.88	5.31	0.98	15.98	0.98
1.08	8.84	0.98	23.68	0.98
1.28	16.77	0.98	37.71	0.98

The reduced moduli were then calculated from intrinsic moduli by taking molecular weights and temperature into account. The longest relaxation time was calculated by using the Rouse approximation:

$$\tau = \frac{6[\eta]\,\eta_s\,M}{\pi^2\;\;R\,T}$$

where $[\eta]$ is the intrinsic viscosity, η is the solvent viscosity, M is the molecular weight and T is the absolute temperature.

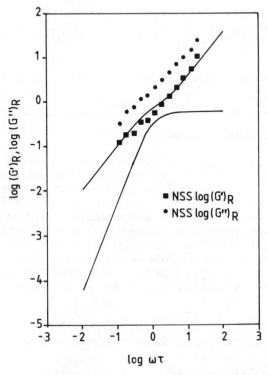

Figure 6   Comparison of experimental reduced moduli of NSS with rod model

The fit of the experimental reduced moduli $[G']_R$ and $[G'']_R$ with the theoretical rigid rod model of Marvin and McKinney[18] for apple pectin of degree of methylation 73.5% is shown in Figure 6. The figure clearly shows that this apple pectin does not follow rod-like behaviour. Experimental reduced moduli were also compared with predictions of the Rouse and Zimm models. The comparison of the reduced moduli with predictions of the Rouse theory is shown in Figure 7. It can be seen from the figure that the fit improves but compared to the rod-like model but nevertheless the data is not well approximated by this theory either. The comparison of the experimental reduced moduli with the predictions of theoretical random coil theory of Zimm are shown in Figure 8 for 73.5% pectin. The data follow the expected values for Zimm type behaviour relatively well with the limiting slope of the reduced moduli closer to the calculated slopes for the Zimm model of 2/3 than those of Rouse of 1/2. This suggests a certain level of intermolecular interaction. This interaction is expected since like charges on the molecule will tend to repulse and opposite charges will tend to attract in essence providing an environment for considerable intermolecular interaction.

The comparison between experimental reduced moduli and the predictions coming from the Zimm model clearly shows that apple pectin with 40.9% degree of esterification also follows Zimm type behaviour. This is shown in Figure 9. This result is also consistent with structural expectations about the molecule. A decrease in degree of esterification in effect releases more charged groups which are available for intermolecular interaction.

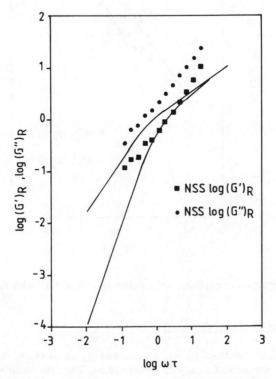

<u>Figure 7</u>   Comparison of experimental reduced moduli of NSS with Rouse model

The reduced moduli for four apple pectins in the middle range of degree of esterification were also compared with the theoretical rod-like and random coil theories of Rouse and Zimm and show similar results which suggest random coil configuration of Zimm type as well[20]. The results obtained with dilute solutions of apple pectins find no evidence that apple pectins show long range rigidity. Studies with citrus pectins[15] and with tomato pectins[29] also showed that both pectin types of varying degree of esterification followed Zimm type behaviour.

<u>Concentrated Systems.</u>   The Bird-Carreau model has been tested with a wide variety of food materials including guar and carrageenan gums, carboxymethyl cellulose, high moisture wheat gluten and wheat glutenin: hard and soft wheat flour as well as mixtures of guar gum and sodium carboxymethylcellulose. The following discussion will present data for some of these materials.

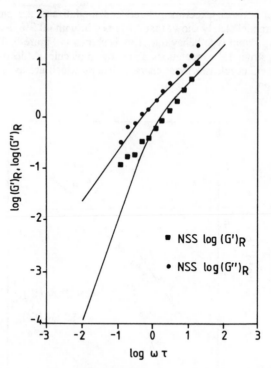

<u>Figure 8</u>   Comparison of experimental reduced moduli of NSS with Zimm model

Determination of Empirical Constants

The empirical constants η_0, λ_1, λ_2, α_1, and α_2 are obtained from geometrical constructions composed of steady or dynamic data. The zero shear viscosity η_0 was obtained from steady shear viscosity versus shear rate data. The value of α_1 is determined from the power law region of the viscosity. The slope of this linear region is equal to

$$\frac{1-\alpha_1}{\alpha_1}$$

The value of λ_1 is the longest relaxation time obtained from steady shear data. The determination of α_2 is from dynamic viscosity data. The power law region of the dynamic viscosity is equal to $(1 - \alpha_1)/\alpha_2$. This constant can also be determined from η''/ω data where the slope of the power law region is equal to

$$\frac{1-\alpha_1-\alpha_2}{\alpha_2}.$$

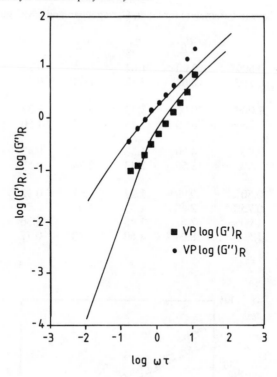

<u>Figure 9</u>   Comparison of experimental reduced moduli of VP with Zimm model

λ_2 can be determined from η' data in a similar manner as λ_1 from η data. The instrumental limitation of a low frequency limit of 0.01 s^{-1} means that the region where η' tends to a limiting value usually cannot be reached with most instruments. Therefore a λ_2 value which gives a satisfactory fit for ψ_1, η', η''/ω predictions when compared to experimental data is often used in the simulation of the viscoelastic properties.

<u>Application of the Bird-Carreau Model to Concentrated Gum Dispersions</u>

Concentrated solutions of polysaccharides in the concentration range of value to the food industry have been shown to be non-Newtonian in steady shear flows at large enough shear rates[30,31]. In transient flows, such as small amplitude oscillatory shear flows, they show both elasticity and viscosity and are linear viscoelastic at small strains.

Rheological modelling is ideal when a single model can explain data obtained in several kinds of experiments. Then data obtained from one kind of experiment can be used to predict data which would have been obtained from another kind of experiment. The Bird-Carreau model is able to satisfy this need.

The empirical constants η_0, λ_1, λ_2, α_1 and α_2 for 1% and 1.25% guar carrageenan and CMC dispersions are shown in Table 3.

Table 3   Bird-Carreau parameters for Guar, Carrageenan and CMC gums

		η_0	α_1	α_2	$\lambda_1(s)$	$\lambda_2(s)$
Guar	0.50%	2.00	1.62	1.04	0.21	0.40
	0.75%	12.62	2.12	1.53	0.42	0.87
Guar	1.00%	20.00	3.33	3.19	1.33	2.50
	1.25%	70.00	5.85	5.67	1.56	3.57
Carra-geenan	2.0%	4.80	1.88	1.81	0.43	0.44
	2.5%	13.50	2.43	2.24	0.59	0.76
CMC	0.50%	0.99	1.27	0.01	0.12	0.21
	0.75%	2.60	1.33	1.02	0.26	0.22
	1.00%	7.61	1.45	1.05	0.36	0.33
	1.50%	28.94	1.60	1.30	0.77	1.54

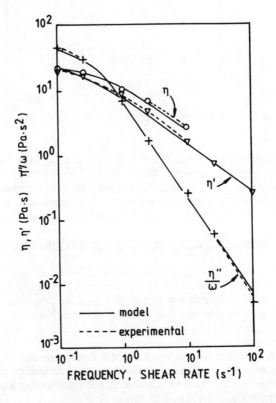

Figure 10 Predictions of the Bird-Carreau model for 1.0% guar: solid lines represent experimental data and dotted lines represent predictions of the Bird-Carreau model

In Figure 10, predictions using the Bird-Carreau model are compared with experimental data for 1.0% guar solution in the frequency/shear rate range of 0.1 - 100 s^{-1}. For η''/ω the model very accurately predicted the high frequency region. The coefficient of determination (R^2) in this region was 0.99. The low frequency region was somewhat less accurately predicted. The R^2 in this region was 0.97. This could be due to the fact that parameters λ_1, λ_2, α_1 and α_2 were determined using the procedure of Bird et al.[2] previously described. This procedure was selected because of its simplicity. An alternative procedure could be used to curve-fit the η vs. $\dot{\gamma}$ and η' vs. ω data to obtain the four parameters. Similarly, experimental values of η' vs. ω and η vs. $\dot{\gamma}$ were fitted quite accurately by the model as demonstrated by R^2 values of 0.99 in high frequency region and R^2 values of 0.98 in the low frequency region.

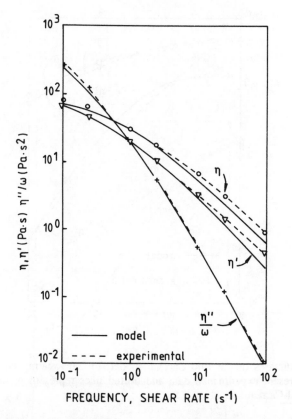

Figure 11 Predictions of the Bird-Carreau model for 1.25% guar: solid lines represent experimental data and dotted lines represent prediction of the Bird-Carreau model

Results for 1.25% guar are shown in Figure 11. The Bird-Carreau model predicts η''/ω vs. ω quite well at both high and low frequency regions. The R^2 of the fit is 0.98 in both regions. Similarly the low and high shear rate/frequency regions of η and η' are well simulated with R^2 values of 0.98.

The 2.0% and 2.5% dispersions of carrageenan were modelled in the same manner as the guar dispersions shown in Figures 12 and 13. η''/ω values are closely predicted in the moderate to high frequency region for both concentrations with R^2 values of 0.99 in all cases. At low frequencies the model predicts a plateau region which was not observed experimentally at either concentration. When the low frequency region was included in goodness of fit estimates the R^2 dropped to 0.93 for 2% carrageenan and to 0.95 for 2.5% carrageenan. Experimental values of η vs. γ and η' vs. ω were also fairly well predicted for the 2.0% carrageenan dispersions. The overall R^2 was 0.98 for both η and η'. For the 2.5% carrageenan dispersions η and η' were also predicted well with R^2 of 0.98 in each case.

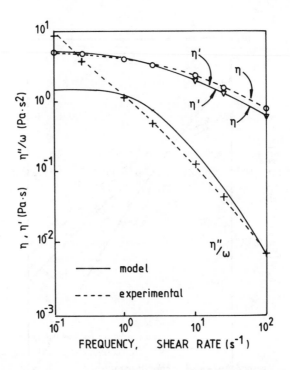

Figure 12 Predictions of the Bird-Carreau model for 2.0% carrageenan: solid lines represent experimental data and dotted lines represent predictions of the Bird-Carreau model

Blends of CMC and Guar Gum

Due to functionality and cost consideration blends of food gums are often used in food formulations; therefore, the ability to predict steady and oscillatory rheological properties of blends of gums from component properties becomes a significant capability.

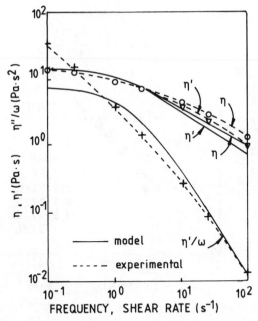

Figure 13 Predictions of the Bird-Carreau model for 2.5% carrageenan: solid lines represent experimental data and dotted lines represent predictions of the Bird-Carreau model

A generalized correlation to predict rheological constants from concentration and molecular weight of the following form was used:

$$f(c_{blend}, \overline{M}_{w,blend}) = p_0 (c_{blend})^{p1} (\overline{M}_{w,blend})^{p2} \qquad (*)$$

where p_0, p_1, p_2 = parameters to be determined;
  c = concentration (g/100 mL);
  M_w = weight-average molecular weight;
  $f(c_{blend}, M_{w,blend})$ = $\eta_0, \lambda_1, \lambda_2, s_\eta$;
  $s_\eta = -slope_\eta$; and $s_{\eta'} = -slope_{\eta'}$

and

$$\overline{M}_{w,blend} = X_1 \overline{M}_{w,1} + X_2 \overline{M}_{w,2}, \qquad X_i \; = \text{mass fraction}$$

$$c_{blend} \;\;\; = v_1 c_1 + v_2 c_2, \qquad v_i \; = \text{volume fraction}$$

The least squares estimators of best fit p_0, p_1, p_2 were determined for experimental values of η_0, λ_1, λ_2, S_η and $S_{\eta'}$ and substituted into equation (*) to calculate predicted values. Predicted values of η_0, λ_1, λ_2, α_1, and α_2 are presented in Table 4. Plot of experimental vs. predicted values of η_0 and λ_1 are presented in Figures 14 and 15. Values for η_0 are plotted on log-log coordinates to accommodate the range of magni-

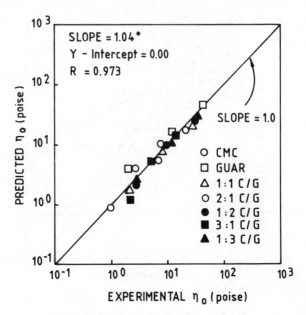

<u>Figure 14</u>   Experimental η_0 vs. predicted η_0

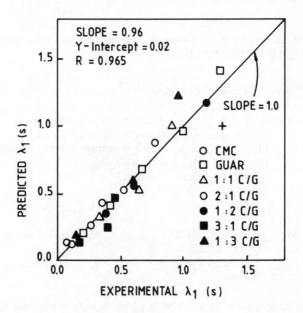

<u>Figure 15</u>   Experimental λ_1 vs. predicted λ_1

tudes; however, reported magnitudes of slope, Y-intercept and R values are for linear coordinates. The prediction of λ_1, λ_2, S_η and $S_{\eta'}$ based on C_{blend} and $M_{w,blend}$ worked quite well[23].

Table 4   Bird-Carreau constants predicted from c_{blend} and $M_{w,blend}$

Material[a]	Conc. (%)	η_0 (poise)	λ_1 (s)	λ_2 (s)	α_1	α_2
CMC	0.50	0.91	0.13	0.20	1.31	0.92
	0.75	3.96	0.26	0.40	1.41	1.07
	1.00	11.24	0.43	0.66	1.50	1.20
	1.50	48.93	0.88	1.33	1.70	1.46
3:1	0.50	1.24	0.14	0.24	1.37	0.97
	0.75	5.41	0.29	0.48	1.50	1.14
	1.00	15.35	0.48	0.78	1.62	1.30
	1.50	66.84	0.97	1.56	1.88	1.63
2:1	0.50	1.39	0.15	0.25	1.40	0.98
	0.75	6.05	0.30	0.50	1.53	1.61
	1.00	17.17	0.49	0.83	1.67	1.34
	1.50	74.75	1.01	1.66	1.97	1.70
1:1	0.50	1.76	0.16	0.28	1.46	1.03
	0.75	7.67	0.32	0.57	1.63	1.23
	1.00	21.76	0.53	0.94	1.80	1.44
	1.50	94.77	1.08	1.88	2.20	1.89
1:2	0.50	2.28	0.17	0.33	1.54	1.08
	0.75	9.93	0.35	0.65	1.75	1.33
	1.00	28.18	0.58	1.07	1.99	1.59
	1.50	122.71	1.18	2.15	2.55	2.20
1:3	0.50	2.62	0.18	0.35	1.59	1.12
	0.75	11.40	0.36	0.70	1.84	1.39
	1.00	32.36	0.60	1.15	2.11	1.69
	1.50	140.91	1.23	2.31	2.82	2.43
Guar	0.50	4.14	0.21	0.45	1.83	1.28
	0.75	18.03	0.42	0.89	2.25	1.70
	1.00	51.18	0.70	1.46	2.79	2.22
	1.50	222.85	1.42	2.93	4.66	4.01

[a] Blends are CMC/guar

Regression analysis of the experimental data with concentration (c) and M_w data resulted in an empirical equation capable of predicting each material constant. These equations are as follows for CMC, guar gum and CMC/guar blends of ratios 3:1, 2:1, 1:1, 1:2 and 1:3 by weight in the concentration range of 0.5 to 1.5% by weight:

$$\eta_p = 1.06 \cdot 10^{19} \qquad c_{blend}^{3.63} \overline{M}_{w, blend}^{-2.94}$$

$$\lambda_1 = 1.81 \cdot 10^{5} \qquad c_{blend}^{1.75} \overline{M}_{w, blend}^{-0.92}$$

$$\lambda_2 = 1.63 \cdot 10^{9} \qquad c_{blend}^{1.72} \overline{M}_{w, blend}^{-1.72}$$

$$S_{\eta} = 1.63 \cdot 10^{7} \qquad c_{blend}^{0.50} \overline{M}_{w, blend}^{-1.26}$$

$$S_{\eta'} = 2.17 \cdot 10^{7} c \qquad c_{blend}^{0.32} \overline{M}_{w, blend}^{-1.26}$$

The Bird-Carreau predictions of η, η' and η''/ω are presented in Figure 16 for 1.0% 3:1 CMC/guar blend as an example. Experimental data is superimposed on these plots to judge the aptness of the model. The steady shear viscosity η and the dynamic viscosity η' are extremely well predicted in the shear rate range of 0.1 to 100 s^{-1}. The experimental data as well as the theoretical predictions portray commonly observed behaviour by polymeric dispersions. In this instance η is larger than η' and the model predicts precisely that. η and η' for this blend ratio tend to the same value, a property suggested by the Bird-Carreau model at low shear rates[12].

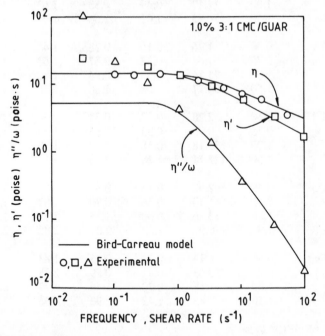

Figure 16 Predictions of the Bird-Carreau model based on concentration and molecular weight for 1.0% 3:1 CMC/Guar blend

η''/ω on the other hand demonstrates similar behaviour to carrageenan gum. That is, the data does not seem to tend to a zero shear constant value. The Bird-Carreau prediction, however, does tend to a zero shear constant value. This results in the ability of the prediction to succeed only in the frequency range of 1 to $100s^{-1}$. However, the discrepancy between the experimental data and the theoretical model widens as the shear rate tends to zero. In this case the smallest shear rate where measurements could be made was 10^{-2} s^{-1}.

Simulation of Wheat Glutenin and Gluten Rheological Properties

In the case of cereal biopolymers the rheological properties at moderate to low moisture content are highly significant. Proteins exist in any amorphous metastable glassy state which is very sensitive to changes in moisture, temperature and processing history.

Water acts as a plasticizer of the amorphous regions present depressing the glass transition temperature. Hoseney[32] and Cocero and Kokini[33] showed that both gluten and its high molecular weight component glutenin were glassy, amorphous and plasticizable polymers with water acting as a plasticizer to depress gluten's glass transition temperature.

At temperatures below the glass transition (glassy-state), the apparent viscosity, is successfully predicted by the Arrhenius equation. At temperatures above the glass transition (typically from Tg to Tg+100°C), in the rubbery range, the dependence of viscoelastic properties on temperature is successfully predicted by the Williams, Landel and Ferry[34] equation . Each of these regions has implications in terms of dough processing. During extrusion, dough is expected to form and flow freely while after extrusion it is expected to be in the glassy region to generate a crisp texture. Similarly during sheeting, dough needs to be viscoelastic to be shaped and formed adquately while after baking many baked products need to be crispy and crunchy which requires that the material be in the glassy state.

Prediction of rheological properties is a key issue for product and process design. In the "free-flow" regions where the material flows the Bird-Carreau model should be able to predict the material rheological of both gluten and glutenin.

The glass transition line in the state diagram for glutenin was obtained from experiments conducted in the moisture range of 0%w-28%w, with Tg ranging between 141.9°C to -71.7°C. The moisture and temperature domain below this line corresponds to the rubbery domain. The Gordon and Taylors'[35] equation was fitted to the glass transition data to give a K value of 6.62 Figure 17. From the annealing experiments the maximum concentrated solution (Tg') region was obtained. This horizontal line depicts the region in the state diagram where ice and rubber coexist as shown in Figure 17.

When the time-temperature superposition principle was applied for the 40%w glutenin using 25°C as the reference temperature, the data was nicely superimposed to obtain the high shear rate viscosity data shown in Figure 18. The shift factor (A_t) temperature dependence or the temperature dependence of the apparent viscosity for 40%w glutenin was better predicted by the Arrhenius equation. Arrhenius equation better predicted the temperature dependence of the apparent viscosity of 40%w glutenin, suggesting that we are in the domain of moisture and temperature where the material flows. The low shear rate viscosity data obtained using the mechanical spectrometer (Rheometrics RMS-800) and Rheometrics stress rheometer (RSR) superim-

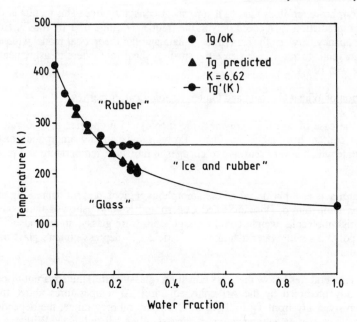

<u>Figure 17</u> Gordon and Taylor's fit for the glass transition in wheat glutenin state diagram

posed very nicely with the viscosity data obtained from the capillary rheometer (Instron) providing steady viscosity data ranging from 10^{-6} to 10^5 1/s Figure 18.

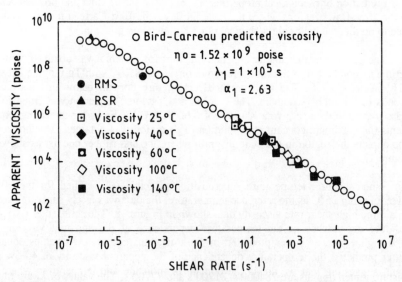

<u>Figure 18</u> Bird-Carreau prediction of the steady viscosity for 40% moisture glutenin at 25°C

The Bird-Carreau equation successfully predicted the apparent steady shear viscosity for 40% moisture glutenin at 25°C . The Bird-Carreau parameters, zero-shear viscosity η_0, α_1 and λ_1 were obtained from the master or composite curve with values of $1.52 \cdot 10^9$ poise, 2.628 poise $\cdot$ s, $1 \cdot 10^5$ s, respectively. These results suggest that 40% moisture glutenin is indeed in the "free-flow region" of glutenin's state diagram. In the case of glutenin the principal protein component of wheat flour dough, the presence of disulfide bonds and noncovalent interactions determine the density of entanglements. 40% moisture glutenin at 25°C, experienced rubbery flow where the entanglements slip so that configurational rearrangements of segments separated by entanglements can take place[18].

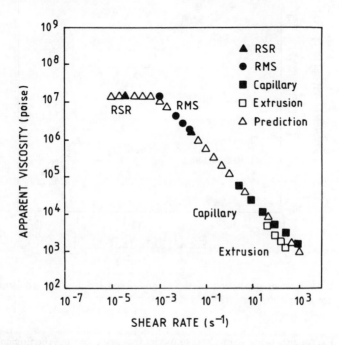

<u>Figure 19</u> Bird-Carreau prediction of the steady viscosity for 52.5% moisture gluten dough at 25°C

Steady viscosity data for gluten with 52.5%, 55%, and 57.5% moisture obtained from RMS, capillary rheometer and stress rheometer are shown in Figure 19 through 21. Three parameters of the Bird-Carreau model, zero shear viscosities (η_0), α_1 and λ_1, of 52.5%, 55% and 57.5% moisture gluten doughs are shown in Table 5. The result shows that these three parameters are moisture content dependent, that the higher the moisture content the larger η_0 and λ_1 and the smaller α_1. α_1 of 55% moisture gluten dough was pretty identical to the one of 57.5% moisture gluten but smaller than that of 52.5% moisture gluten dough. α_2 of 52.5%, 55% and 57.5% moisture gluten doughs showed the same dependence on moisture content as α_1. λ_2 values were computed by iterative calculations. The calculated λ_2 values of 52.5%, 55% and 57.5% moisture gluten doughs are 284000 s, 81700 s and 77700 s. The values of λ_1 are much smaller than the values of λ_2 in all gluten doughs with different moisture content implies that the gluten is much easier to form a network than to destroy a network.

Using these parameters in the Bird-Carreau model gives excellent prediction for steady viscosity, η' and η''/ω for all three moisture gluten doughs we studied in this project and the R^2s are larger than 0.99. An example of this fit for gluten doughs is given in Figure 22.

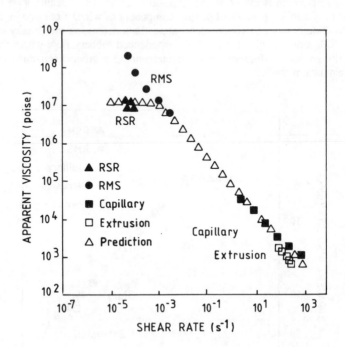

<u>Figure 20</u> Bird-Carreau prediction of the steady viscosity for 55% moisture gluten dough at 25°C

<u>Table 5</u>   Parameters using Bird-Carreau model for predicting the rheological properties of gluten doughs with different moisture contents

Moisture (%)	η_0(poise)	λ_1(s)	α_1	α_2	λ_2(s)
52.5	$1.104 \cdot 10^8$	14040	3.11847	3.32174	$7.174 \cdot 10^6$
55.0	$8.818 \cdot 10^7$	7870	3.52336	3.87928	$4.270 \cdot 10^6$
57.5	$4.943 \cdot 10^7$	3380	3.50869	3.52602	$7.050 \cdot 10^5$

<u>Simulation of Wheat Flour Dough Rheological Properties</u>

Wheat dough are highly viscoelastic materials and their rheology affects processing dramatically. For example during sheeting both steady shear and steady uniaxial extensional properties affect processing performance of the dough. While the Bird-Carreau model is able to predict steady shear properties it is not able to predict steady uniaxial properties. Alternative constitutive models are considered[36] for the latter.

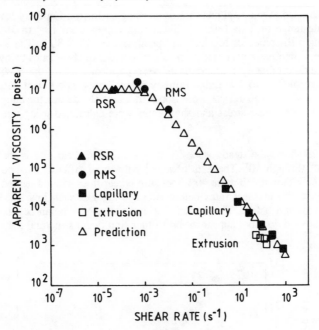

Figure 21 Bird-Carreau prediction of the steady viscosity for 57.5% moisture gluten dough at 25°C

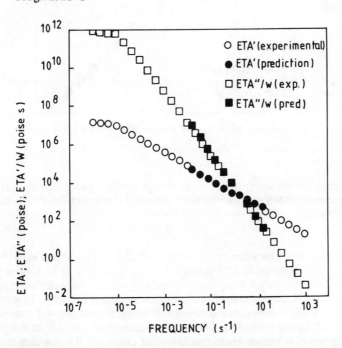

Figure 22 Bird-Carreau prediction of the η' and η''/ω for 55% moisture gluten dough at 25°C

We conducted rheological measurements using an Instron Capillary Rheometer in the shear rate range of 2 to 1800 s^{-1}. Shear rate ranges from 10^{-4} to 10^{-1} s^{-1} were covered using a Rheometrics Mechanical Spectrometer (RMS-800). The Rheometrics constant stress rheometer was used to obtain viscosities at shear rates below 10^{-4} s^{-1}. Small amplitude oscillatory measurements were performed using the RMS and covered the range of 0.01 to 100 rad/s. Primary normal stress coefficient data for shear rates in the range of 9 to 60 s^{-1} were obtained using slit rheometry coupled with a Brabender extruder shear stress and shear rates were calculated using the method of Han[37].

Figure 23 shows the steady viscosity as a function of shear rate in the shear rate range of 10^{-6} through 10^3. The data obtained with three different methods superimpose very nicely showing that the 40% moisture wheat flour studied tended towards a zero shear viscosity at shear rates below 10^{-5} s^{-1}. Data at such low shear rates has never been reported before. This material becomes non-Newtonian beyond the shear rate of 10^{-5} s^{-1} and does not appear to show other significant structural transitions.

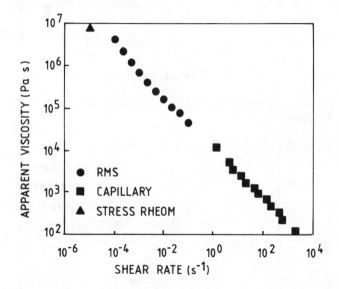

Figure 23 Apparent viscosity (η) vs. shear rate(γ) obtained using a mechanical spectrometer (RMS), a capillary rheometer, and a constant stress rheometer for 40%

Figure 24 shows the primary normal stress coefficient ψ_1 as a function of shear rate in the shear rate range of 10^{-4} through 10^2 s^{-1}. Only data up to a shear rate of $5 \cdot 10^{-2}$ s^{-1} could be obtained using the RMS and only data in the shear rate range of 5 through 80 s^{-1} could be obtained using slit rheometry during extrusion. However, the data appeared to follow power-law behaviour. No zero shear trend was observed. To simulate the rheological properties of this 40% moisture hard wheat flour dough using the Bird-Carreau constitutive model the following values for the empirical constants were used.

$$\eta_0 = 8 \cdot 10^6 \text{ Pa s},$$

$$\alpha_1 = 2.4942,$$

$$\lambda_1 = 50\ 000 \text{ s},$$

$$\alpha_2 = 1.8373,$$

$$\lambda_2 = 3800 \text{ s}.$$

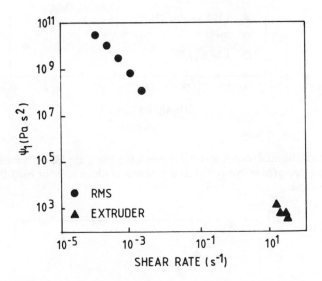

Figure 24 Primary normal stress coefficient(ψ_1) vs. shear rate (γ) using a mechanical spectrometer and slit die rheometry using a single extruder

The experimental and predicted values of the steady viscosity function versus shear rate are given in Figure 25. The experimental data are well simulated by the Bird-Carreau model. Comparison of experimental data versus the predicted data shows a high degree of superposition throughout the range of viscosities.

The prediction of the primary normal stress coefficient shown in Figure 26 is fairly close considering the difficulty in experimental determination of N_1, hence ψ_1. Experimental data at the low shear rates obtained using the mechanical spectrometer show a significant discrepancy relative to predicted values. The source of this discrepancy might be in the experimental difficulties in obtaining reliable N_1 measurements. Using empirical constants obtained from other viscoelastic functions, the Bird-Carreau model provided a reasonably accurate prediction of ψ_1.

The prediction of η' and the experimental data versus frequency in Figure 27 show that the prediction works well. This is especially so at the intermedite frequencies where the prediction is very close. The experimental data are seen to deviate slightly at the extremes of the frequency range.

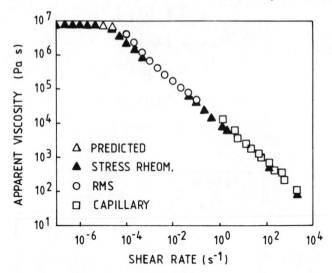

Figure 25 Experimental values and those predicted using the Bird-Carreau model of the apparent viscosity (η) as a function of shear rate for hard flour dough sample

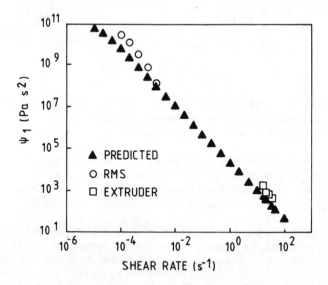

Figure 26 Experimental values and those predicted using the Bird-Carreau model of the primary normal stress coefficient(ψ_1) as a function of shear rate for 40% moisture

Figure 28 shows the predicted and experimental values of η''/ω versus frequency. In the range of frequencies tested there is a very high degree of superposition of the

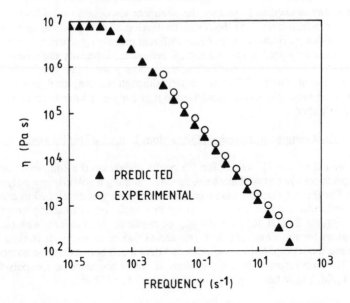

Figure 27 Experimental and predicted values of the dynamic viscosity (η')vs. frequency for the 40% moisture hard flour dough sample

experimental and predicted data. The performance of this prediction is better than those of the other functions.

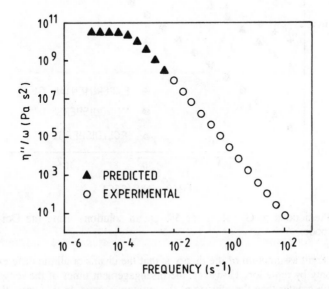

Figure 28 Experimental and predicted values of η''/ω vs. frequency for the 40% moisture hard wheat flour dough

The ability to accurately predict the nonlinear viscoelastic behaviour of a wheat dough is of practical interest to the scientists developing new products or technologies in the food industry. Because of the many different processing schemes in use it is necessary to accurately predict the rheological behaviour, especially the steady viscosity and the primary normal stress coefficient, throughout a shear rate range which is relevant to processing. The Bird-Carreau model, although semiempirical, provides the accuracy and the versatility which should make it of particular interest to those working with wheat doughs.

Predicting Concentrated Apple Pectin Dispersions Using the Doi-Edwards Model

To predict G' and G" values for 5% pectin dispersion the equations assuming a monodisperse polymer as well as the equations assuming a polydisperse polymer were used[26,28]. For the polydisperse case a computer program was developed to account for a small molecular fraction measured using low angle light scattering coupled with HPLC[26]. Figures 29 and 30 show the plot of predicted values along with the experimental values for the simulation of $G'(\omega)$ and $G''(\omega)$ respectively. It is clear that the polydisperse model explains the experimental data much better than the monodisperse model. This is no surprise since apple pectin is highly polydisperse and polydispersity ratios (M_w/M_n) have been reported ranging from 15 to 45[26,38].

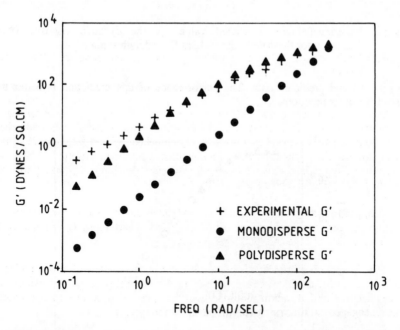

Figure 29 Predictions of G' values for 5% pectin solution using the Doi-Edwards model

An inherent assumption of the theory is that the chains of all the different lengths can move only by reptation. However, the disengagement times of the longest chains may be much greater than the lifetime of the temporary cage. In that case, the longest chains may diffuse not only by reptation but randomly directed segmental jumps. On the other hand, very short chains may diffuse laterally and not contribute to reptation.

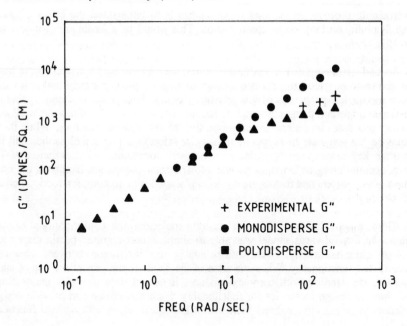

<u>Figure 30</u> Predictions of G" values for 5% pectin solution using the Doi-Edwards model

Another factor which may influence reptation is branching. Apple pectins are constructed of homogalactouronan and rhamnogalactouronan regions with side chains of arabinogalactan. These side chains are highly branched[39]. Short side branches (much shorter than the chain length between entanglement points) are expected to have negligible effect upon reptation, while long and extensive branching may modify the characteristics of the cage and tube substantially.

Also, pectin is an anionic molecule i.e. the chains are in a more extended form as the same charges repel each others. These extended, much longer chains can modify the characteristics of cage and tube considerably.

Here, in apple pectin, which is high molecular mass sample with broad molecular weight distribution, chains of all sizes are compelled to diffuse only by reptation, and yet the cage lifetimes, a communal property of the system may be very small compared to the disengagement time of the largest chains in the system.

<u>Concluding Remarks</u>

In conclusion, constitutive models have been shown to be useful in many ways.

First, dilute solution theories have been shown to be useful to characterize long range conformation and flexibility of carbohydrate as well as protein based polymers. Rheological measurements and constitute theories predicting rheological properties compliment short range studies such as neutron scattering. One important advantage of combining short range vs. long range studies is to understand the effect of short range flexibility on long range conformation. This would be a fertile area for new research.

Second, semi-empirical constitutive models such as the Bird-Carreau model have been shown to be useful to generate a database able to predict steady and small amplitude rheological properties of viscoelastic materials. This database should enable us to estimate important parameters such as recoverable strain which affects die swell and recovery processes in general in the processing of viscoelastic materials. While these models do not originate from specific molecular information nevertheless they still incorporate key assumptions pertaining to network formation and dissolution which clearly occurs during deformation processes. The semi-empiricism facilitates the estimation of parameters and makes the model easily applicable to complex food materials such as doughs.

Third, a constitutive model with accurate molecular and conformational origins such as the Doi-Edwards model was able to simulate concentrated pectin dispersion rheology quite accurately. Such studies enable the correlation between chemical structure and functionality. While the state of the art does not permit the use of such detailed understanding with complex mixtures it nevertheless provides major clues about how to design molecules for functionality. Such knowledge can provide design guidelines to genetically engineer materials of biological origin with optimal functionality.

Clearly there is need for much more work in this area of food rheology. It is first necessary to bring into the field constitutive models able to predict extensional properties as well. In addition, constitute models which can predict disperse system (suspensions, emulsions) rheology need to be applied or developed. Progress has to be made in numerical methods so that non-linear constitutive models can be incorporated into process design. Finally, new constitutive models need to be developed which will have the capability to incorporate the diverse and complex structural properties of foods. The predictive capability developed through all this effort will result in product design and process improvements rules leading to improved food products for the consumer.

2 ACKNOWLEDGEMENTS

This is paper D-10544-20-92 of New Jersey Agricultural Experiment Station, supported by State Funds and the Center for Advanced Food Technology (CAFT). The Center for Advanced Food Technology is a New Jersey Commission on Science and Technology Center. This work was also supported by the National Science Foundation (NSF), the Campbell Soup Company and M&M Mars, a division of Mars Incorporated.

3 REFERENCES

1. A. Einstein, Ann. Physik, 1906, 19, 289.
2. R.B. Bird, R.C. Armstrong and O. Hassager, 'Dynamics of Polymeric Liquids', John Wiley & Sons, New York, 1977, Vol. 1.
3. R.B. Bird et al., 'Dynamics of Polymeric Liquids', John Wiley & Sons, New York, 1987, Vol. 1.

4. E.B. Bagley, 'Food Extrusion Science and Technology', Marcel Dekker, New York, 1992, Chapter 13, pp. 203.
5. R. Darby, 'Viscoelastic Fluids', Marcel Dekker, New York, 1976.
6. A.H. Bloksma, Rheol. Acta, 1962, 2, 217.
7. P. Sherman, 'The Encyclopedia of Emulsion Science', Marcel Dekker, New York, 1983, Vol. 1, Chapter 3.
8. E.R. Fischbach and J.L. Kokini, J. Food Science, 1987, 52, 1748.
9. A.R. Carrillo and J.L. Kokini, J. Food Science, 1988, 53, 1352.
10. G. Yilmazer, A.R. Carrillo and J.L. Kokini, J. Food Science, 1991, 56, 513.
11. E.B. Bagley and D.D. Christianson, J. Rheol., 1987, 31, 405.
12. J.L. Kokini, K. Bistany and P. Mills, J. Food Science, 1984, 49, 1569.
13. J.L. Kokini and G.J. Plutchok, Food Technology, 1987, 41, 89.
14. S.J. Dus and J.L. Kokini, J. Rheol., 1990, 34, 1069.
15. T.C. Chou, N. Pintauro and J.L. Kokini, J. Food Science, 1991, 56, 1365.
16. P.E. Rouse Jr., J. Chem. Phys., 1953, 21, 1272.
17. B.H. Zimm, J. Chem. Phys., 1956, 24, 269.
18. R.S. Marvin and J.E. McKinney, 'Physical Acoustice', (Ed. W.P. Mason) Academic Press, New York, 1965, Vol. B.
19. J.D. Ferry, 'Viscoelastic Properties of Polymers', John Wiley & Sons, New York, 1980.
20. J.L. Kokini, T.C. Chou and J.H. Hwang, J. Rheol., 1993, in review.
21. K.L. Bistany and J.L. Kokini, J. Rheol., 1983, 27, 605.
22. H.M. James, J. Chem Phys., 1947, 15, 651.
23. G. Plutchok and J.L. Kokini, J. Food Science, 1986, 51, 1284.
24. W.R. Leppard, Ph.D. Thesis, University of Utah, 1975.
25. E. Doi and R.H. Edwards, J. Chem. Soc., Part I, 1978, Faraday Trans II, 74, 1789.
26. S.H. Shrimanker, M.S. Thesis, Rutgers University, 1989.
27. E. Doi and R.H. Edwards, J. Chem. Soc., Part II, 1978, Faraday Trans. II, 74, 1802.
28. Rahalkar et al., J. Rheol., 1985, 29, 955.
29. J.L. Kokini and T.C. Chou, J. Text. Stud., 1993, in press.
30. J. Schurz, 'Proceedings of the 7th International Congress on Rheology', (Eds. C. Klason and J. Kubat), Chalmus University of Technology, Guthenburg, Sweden, 1976, p. 123.
31. M. Glicksman, 'Food Hydrocolloids', (Ed. A. Glicksman), CRC Press, Boca Raton, Florida, 1982, Vol. 1:4.
32. R.C. Hoseney, K. Zeleznak and C.S. Lai, Cereal Chem., 1986, 63, 285.
33. A.M. Cocero and J.L. Kokini, J. Rheol., 1991, 35, 257.
34. M.L. Williams, R.F. Lanche and J.D. Ferry, J. Amer. Chem. Soc., 1955, 77, 3701.
35. M. Gordon and J.S. Taylor, J. Appl. Chem, 1952, 2, 493.
36. M.H. Wagner, Rheol. Acta, 1976, 15, 136.
37. C.D. Han, 'Rheology in Polymer Processing', Academic Press, New York, 1976.
38. T.C. Chou and J.L. Kokini, J. Food Science, 1987, 62, 1658.
39. A.G. deVries, J. Voragen, F.M. Rombouts and W. Pilnik, 'Chemistry and Function of Pectins', American Chemical Society Symposium, Series 310, Washington D.C., 1985, 38.

VISCOELASTIC PROPERTIES OF MIXED POLYSACCHARIDES SYSTEMS

J.L. Doublier, C. Castelain and J. Lefebvre

INRA-LPCM, BP 527, 44026 NANTES CEDEX 03, FRANCE

1 INTRODUCTION

Polysaccharides are extensively used in the food industry because they allow the rheological and textural characteristics of food products to be controlled. In particular, their mixtures can lead to novel or improved properties and furthermore can give rise to significant cost savings. These special properties are more often the result of synergistic interactions. Several polysaccharide mixtures are used in the food industry for these synergistic properties[1]. However, the most widely exploited mixed systems are those involving a galactomannan, locust bean gum. When combined with κ-carrageenan, agar or xanthan, locust bean gum gives systems exhibiting strong gel-like properties[2-6]. This can occur even under conditions in which neither of the polysaccharides would gel alone. More recently, there has been an increasing interest for galactomannans from other botanical sources such as tara gum or for related mannans like glucomannans. These polysaccharides exhibit similar synergistic properties to locust bean gum. Such properties have not been reported in blends with guar gum. However, there is some account of an increase in viscosity in xanthan/guar mixtures[7], suggesting that similar phenomena, and similar interactions, as with locust bean gum can be experienced.

The viscoelastic properties of polysaccharide systems can be easily characterized by oscillatory shear measurements under small-deformation conditions. The shape of mechanical spectra, G' and G" variations as a function of frequency, offers a good means to describe the overall properties and is helpful to understand the underlying physical and chemical mechanisms. The viscoelastic behaviour of individual polysaccharides in aqueous medium is well known. On one hand, non-gelling polysaccharides such as galactomannans (guar gum or locust bean gum) exhibit properties that are typical of polymer solutions. On the other hand, gelling polysaccharides such as alginates, carrageenans (κ or ι) or pectins yield strong gels, the viscoelastic properties of which are classically characterized by a flat frequency dependence of $G'(\omega)$ and $G' >> G"$[8]. Blends of polysaccharides, however, show more complex behaviours that are not related to those of their individual components. It has thus been shown, for instance, that in the case of xanthan/locust bean gum mixtures two plateau regions exist in the mechanical spectrum[4,5].

In the present work, the viscoelastic properties of mixtures of a galactomannan (guar gum and locust bean gum) with xanthan gum are described. Spectacular results are obtained in relation to synergistic interactions. For instance, peculiar viscoelastic behaviours have been found at a polysaccharide/galactomannan ratio as low as 1:99 irrespective of the type of galactomannan. Despite the high content in galactomannan, the properties were very different from those of macromolecular solutions such as displayed by galactomannan alone. Typically, the behaviour was gel-like at low frequency whereas it remained close to that of the individual galactomannan at high frequency. This means that a weak gel has been formed. These results are discussed on the basis of current models that have been proposed for the understanding of the mechanisms involved in synergistic interactions between polysaccharides.

2 EXPERIMENTAL

Materials

Locust bean gum was supplied by Meyhall Chemical (Switzerland). Guar gum was a gift of Sanofi Bio Industrie (France). These samples were protein-free (< 0.2 wt% protein) and had intrinsic viscosities of 1200 and 1450 ml g^{-1} for guar gum and locust bean gum, respectively. Xanthan gum was supplied by Sanofi Bio Industrie (France). Its intrinsic viscosity was 9550 ml g^{-1} in 0.13 mol/dm^3 KCl. Its acetyl and pyruvate contents were 4.43 and 5.83 wt%, respectively, meaning that xanthan molecules are fully acetylated (1 acetyl group every pentasaccharide repeat unit) and 60% pyruvated.

Preparations of Solutions and Mixtures

Guar gum and locust bean gum solutions were first prepared by dispersion in 0.13 mol g^{-1} KCl at 25°C. The dispersion was then heated to 80°C and kept at this temperature for 30 minutes in order to achieve a total solubilization of macromolecules. The xanthan gum solution was prepared by dispersion in KCl (0.13M) at 25°C with strong stirring for 2 hours. Air bubbles entrapped in the solution due to the strong agitation were then removed by centrifugation.

Blends of the galactomannan with xanthan were obtained by mixing appropriate amounts of each polysaccharide solution at 80°C under magnetic stirring. The hot mixtures were then poured into the rheometer and cooled rapidly to 25°C.

Oscillatory Shear Measurements

These measurements were performed using a Rheometrics Fluid Spectrometer (RFS II) with a cone-plate device (cone angle 2.2°, diameter 5 cm). Measurements were performed after a 30 min ageing period. The amplitude of deformation was chosen in order to remain within the linearity limits of viscoelasticity. This could be as high as 0.4 for galactomannan solutions, whereas it had to be lower than 0.1 for xanthan gum and the xanthan/galactomannan mixtures. The frequency range was most often between 0.01 and 100 rad/s but in the case of xanthan gum this scale has been extended down to 0.001 rad/s. Temperature measurement was fixed at 25°C. Experiments at temperatures ranging from 25°C to 75°C were also performed on the mixtures.

3   RESULTS AND DISCUSSION

<u>Viscoelastic Behaviour of Galactomannans</u>

Figure 1 shows the mechanical spectra, G'(ω) and G"(ω), of locust bean gum solutions at three concentrations (2%, 1% and 0.5%). They are comparable to those published for guar gum[9,10] and can be interpreted in a similar way. At low frequency, the loss modulus G" is higher than the storage modulus and both parameters vary sharply with frequency: G" α ω and G' α ω². The behaviour is said to be 'liquid-like'. As the frequency increases, a cross-over of the curves is experienced since G' increases faster than G" beyond which G'>G". This is indeed more clearly seen at the highest concentration. The response of the material beyond the cross-over frequency is more 'solid-like'. Such a behaviour is typical of macromolecular solutions with topological entanglements[11]. Within the frequency range explored, we observe the terminal zone (at low frequency) and the beginning of the plateau zone (at high frequency) of the complete mechanical spectrum. This means that the rheology of galactomannan solutions is mainly governed by the degree of entanglement of individual macromolecules.

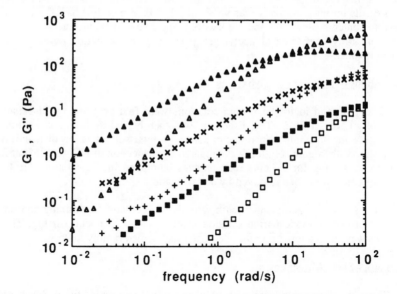

<u>Figure 1</u>   Mechanical spectra of three locust bean gum solutions. Concentrations: 2% (Δ, G'; ▲, G"); 1% (+ ,G'; x , G"); 0.5% (□, G'; ■, G"). Temperature: 25°C

<u>Viscoelastic Behaviour of Xanthan</u>

Figure 2 illustrates the behaviour of xanthan gum at 0.5% as compared to locust bean gum at the same concentration. The difference in viscoelastic properties is clearly evidenced, xanthan showing little frequency dependence of G' and G" in contrast to locust bean gum. Moreover, G' is higher than G" over the whole frequency range investigated (10^{-3} to 100 rad/s). Such a behaviour may appear to correspond to the definition of a weak gel[12,13]. However, a xanthan dispersion flows freely and for this reason can not be considered as a true gel. It has been suggested that this peculiar behaviour is due to associations of ordered chain segments giving rise to a weak three-dimensional network[14,15]. However, another interpretation can be proposed[16]. Despite the 5 decades frequency range covered, the time-scale investigated can still be considered as relatively narrow. The mechanical spectrum displayed may thus be regarded as that of a macromolecular solution with the cross-point of G' and G" occuring at a lower frequency than those accessible. Therefore, the elastic plateau would be observed at a frequency much lower than in the case of galactomannans. That means that the peculiar viscoelastic properties of xanthan are governed by molecular movements with relaxation times significantly longer than for galactomannan solutions. This can simply be ascribed to a lower molecular mobility of xanthan molecules. The dynamic behaviour of xanthan dispersions may thus be related to a higher stiffness of the polysaccharide and its interpretation does not require the formation of a network. Therefore, the main difference in viscoelastic behaviour between galactomannans and xanthan lies in the fact that galactomannans behave as random-coil molecules whereas xanthan chains adopt a helical conformation which is much more rigid.

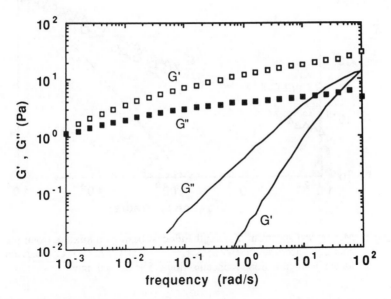

Figure 2 Mechanical spectrum of a xanthan solution (concentration: 0.5%; temperature: 25°C; □, G'; ■, G"). Comparison with locust bean gum at the same concentration (continuous lines)

Behaviour of Mixed Systems at a Low Xanthan:Galactomannan Ratio

Figures 3 and 4 illustrate the properties of mixtures of xanthan with locust bean gum at 0.5wt% total concentration and different xanthan/locust bean gum (X:L) ratios, namely 3:97 in Figure 3, 6:94 and 9:91 in Figure 4. The mechanical spectrum of locust bean gum at the same total concentration is given in Figure 3. A dramatic change in the G' trace was seen while the G" curve was only slightly shifted to higher values. When the xanthan content remained low, X:L= 3:97 and 6:94 in the present examples, the main feature was the flattening of the G' curve towards the low frequency range which assimilated to a plateau and contrasts strongly with the G' variations of locust bean gum. This means that the behaviour is solid-like. Similar results were obtained at any X:L ratio higher than 1:99. Increasing the xanthan content yielded more pronounced effects particularly on the G' plateau values. Moreover as illustrated in Figure 4 for X:L=9:91, the G' trace became more complex and two plateaus were evidenced separated by an inflection point at ~ 0.1 rad s[-1] while the G" curve exhibited a maximum at this frequency. Quite comparable spectra have been reported for xanthan/locust bean gum mixtures[4,5] but at a high X:L ratio (70:30). It is thus clear that the presence of xanthan gum even at a low ratio induces a transition of the system from a macromolecular solution to a structured system with gel-like properties.

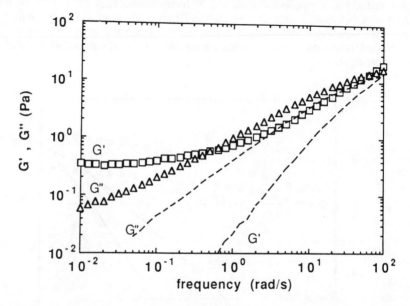

Figure 3   Mechanical spectrum of a xanthan/locust bean gum mixture (total polysaccharide concentration: 0.5%) X:L = 3:97 (□, G'; ∆, G"). Comparison with locust bean gum alone (concentration:0.5%; dashed lines)

Figure 5 provides an illustration of the results obtained with guar gum at xanthan/galactomannan ratios (X:G) of 1:99 and 3:97, compared to guar gum alone. Overall similar tendencies as those seen for xanthan:locust bean gum mixtures were observed but the phenomena appeared somewhat more pronounced. These results were not expected since it is generally believed that guar gum does not develop strong synergistic interactions with xanthan[1]. The fact that we observed qualitatively similar results whatever the type of galactomannan has to be interpreted on a molecular basis. This point will be discussed further.

It is also worth mentioning that such phenomena can be observed when galactomannans are mixed with polysaccharides other than xanthan. An example is given in Figure 6[17]. The system investigated was a κ–carrageenan/locust bean gum mixture at 1 wt% in KCl (0.13 M). The κ–carrageenan/galactomannan ratio was 1/99. Although

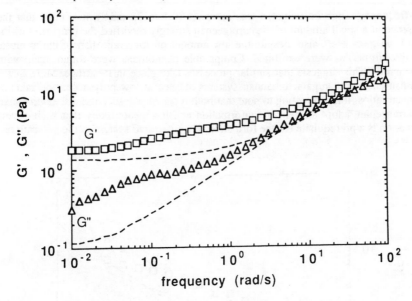

<u>Figure 4</u>   Mechanical spectra of xanthan:locust bean mixtures. Concentration: 0.5%;
(X/L = 6:94; dashed lines; X/L = 9.91, □, G'; Δ, G")

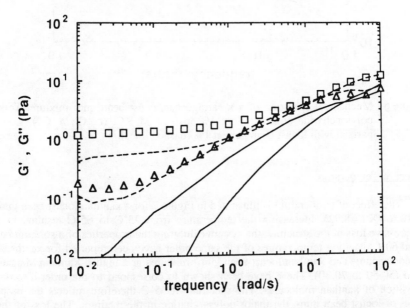

<u>Figure 5</u>   Mechanical spectra of xanthan/guar gum mixtures at two X:G ratios (total
polysaccharide concentration: 0.5%); (X:G= 1:99, dashed lines; X:G=
3:99, □, G'; Δ,G"). Comparison with guar gum (concentration: 0.5%, continuous lines)

the frequency range investigated was narrower (0.1 to 25 rad/s), it is clear that the presence of a small amount of the carrageenan strongly modified the properties of locust bean gum. Here also, despite the low amount of κ-carrageenan in the mixture, gel-like properties were exhibited. Comparable phenomena were found again with guar gum[17.] This suggests that similar processes take place in κ-carrageenan/galactomannan and xanthan/galactomannan systems at least at low polysaccharide/galactomannan ratios. It is interesting to note that both types of polysaccharides, xanthan and κ-carrageenan, adopt an helical conformation and it appears likely that such a conformation is a prerequisite for the formation of a structured system in the present conditions.

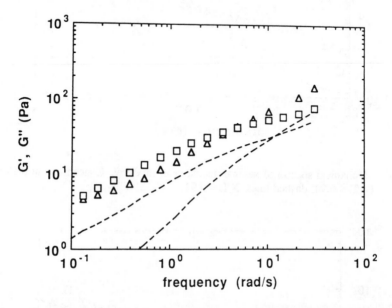

<u>Figure 6</u>   Mechanical spectrum of a κ-carrageenan/locust bean gum mixture. Total polysaccharide concentration: 1% in 0.13 M KCl (□, G'; Δ, G"). Comparison with locust bean gum at 1% (dashed lines)

<u>Effect of Temperature</u>

The effect of temperature is illustrated in Figure 7 for a xanthan/locust bean gum mixture (X:L=8:92). Increasing the temperature from 25°C to 65°C resulted in a progressive loss of the structure, the system exhibiting the properties of a solution beyond 60°C. Melting temperatures of the same order have been reported for xanthan/-glucomannan[18] and xanthan/locust bean gum[5] mixtures at a higher Xanthan/Mannan ratio (50:50 to 70:30). These have been shown to correspond to the order-disorder transition of xanthan molecules. The spectrum at 65°C therefore reflects the properties of locust bean gum, the major polysaccharide in the medium. The loss of the structure thus corresponds to the melting of xanthan molecules which no longer contribute appreciably to the viscoelasticity of the medium. Moreover, these results indicate that, when preparing the mixtures at 80°C in the present study, we allowed xanthan molecules to be in a disordered form before cooling the system. This is

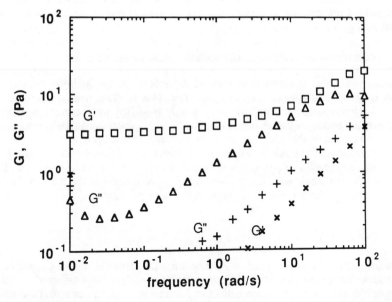

Figure 7 Effect of temperature on viscoelastic properties of a xanthan/locust bean gum mixture (X:L= 8:92). Temperature: 25°C; □, G'; Δ, G". Temperature: 65°C; (x , G'; + , G")

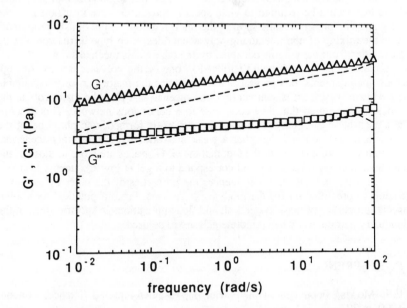

Figure 8 Mechanical spectrum of a xanthan/locust bean gum mixture (total polysaccharide concentration: 0.5%; X:L= 99:1; (Δ, G'; □, G"). Comparison with xanthan gum (dashed lines; concentration: 0.5%)

known to be a prerequisite to develop synergistic interactions in xanthan/locust bean gum mixtures.

Behaviour of Mixed Systems at a High Xanthan/Galactomannan Ratio

It appeared also interesting to look at the effects of the addition of a small amount of galactomannan to a xanthan solution. This is illustrated in Figure 8 for X:L=99:1. Here again, the properties are strongly modified when compared to xanthan alone, the major polysaccharide in the medium. This was particularly evident on G' variations towards the low frequency range. Since locust bean gum is much less viscous than xanthan, it could be expected that replacing xanthan by locust bean gum might have, due to dilution of xanthan, an effect opposite to the one observed. The present result here again can be explained by the structuration of the medium due to the coexistence of the two polysaccharides in the same medium.

4 CONCLUSIONS

The molecular mechanisms underlying the peculiar properties of xanthan/locust bean gum and κ-carrageenan/locust bean gum systems are classically assumed to arise from specific interactions between helices of xanthan or κ-carrageenan and unsubstituted regions of the galactomannan[19]. The fact that guar gum and locust bean gum, at a very low ratio of added xanthan or κ-carrageenan, result in similar properties tends to rule out this interpretation. The main structural difference between the two galactomannans lies in the galactose/mannose ratio and the distribution of galactose along the mannan chain. The length and number of galactose-free regions along the mannan backbone that would be required to yield specific interactions are insufficient in the case of guar. The present results support the interpretation we previously proposed[20] based on the mixing of non interacting polysaccharides, each type of macromolecule being excluded from the volume occupied by the other. This mechanism would result in two separate phases, each being enriched in one of the polysaccharides. At high galactomannan contents, this polysaccharide occupies the major part of the available volume. However, since it is xanthan or κ-carrageenan that play the major role in the properties, it is required that they contribute to a continuous network. This description assumes that a bi-continuous two-phase system is formed, one phase being the galactomannan solution and the other a weak network of the other polysaccharide. This provides an explanation for the fact that the G'-G" traces are close to those of the galactomannan at high frequency and correspond to a gel at low frequency. It is also possible that each polysaccharide influences the properties of the other by leading it to gelation or precipitation for thermodynamic reasons. From a practical viewpoint, the special properties of these systems should find applications in specific areas of the food industry particularly when pourable gels are to be used.

5 REFERENCES

1. E.R. Morris, 'Food gels', (Ed. P. Harris), Elsevier Applied Science, London, 1990, p. 291.
2. P. Cairns, V.J. Morris, M.J. Miles and G.J. Brownsey, Food Hydrocolloids, 1986, 1, 89.
3. P. Cairns, M.J. Miles, V.J. Morris and G.J. Brownsey, Carbohydr. Res., 1987, 160, 411.

4. G. Cuvelier, and B. Launay, 'Gums and stabilizers for the food industry-3', (Eds. G.O.Phillips, D.J. Wedlock and P.A. Williams), Elsevier Applied Science, London, 1986, p. 147.
5. G. Cuvelier, 'PhD Thesis', Université Paris XI, 1988.
6. P.B. Fernandes, M.P. Gonçalves and J.L. Doublier, Carbohydr. Polym., 1991, 16, 253.
7. M. Tako and S. Nakamura, Carbohydr. Res., 1985, 138, 207.
8. A.H. Clark and S.B. Ross-Murphy, Adv. Polym. Sci., 1987, 83, 57.
9. G. Robinson, S.B. Ross-Murphy and E.R. Morris, Carbohydr. Res., 1982, 107, 17.
10. R.K. Richardson and S.B. Ross-Murphy, Int. J. Biol. Macromol., 1987, 9, 257.
11. J.D. Ferry, 'Viscoelastic properties of polymers', J. Wiley and Sons, New York, 1980.
12. R.K. Richardson and S.B. Morris, Int. J. Biol. Macromol., 1987, 9, 257.
13. S.A. Frangou, E.R. Morris, D.A. Rees, R.K. Richardson and S.B. Ross-Murphy, J. Polym. Sci., Polymer Lett. Ed., 1982, 20, 531.
14. E.R. Morris, 'Frontiers in Carbohydrate Research: Food Applications', (Eds. R.P. Millane, J.N. BeMiller and R. Chandrasekan), Elsevier Appl. Sci., London, 1989, p. 132.
15. S.B. Ross-Murphy, V.J. Morris and E.R. Morris, Faraday Symp. Chem. Soc., 1983, 18, 115.
16. J.L. Doublier, B. Launay and B. Cuvelier, 'Viscoelastic Properties of Food Gels', (Eds. M.A. Rao and J. Steffe), Elsevier Applied Science, in press.
17. P.B. Fernandes, 'PhD Thesis', Universidade Catolica Portuguesa, Porto, 1992.
18. P.A. Williams, S.M. Clegg, D.H. Day and G.O. Phillips, 'Food Polymers, Gels and Colloids', (Ed. E. Dickinson), RCS Special Publication No. 82, Cambridge, 1991, p. 339.
19. I.C.M. Dea, E.R. Morris, D.A. Rees, E.J. Welsh, H.A. Barnes and J. Price, Carbohydr. Res., 1977, 57, 249.
20. J.L. Doublier and G. Llamas, 'Food Polymers, Gels and Colloids', (Ed. E. Dickinson), RCS Special Publication No. 82, Cambridge, 1991, p. 349.

DETERMINATION OF THE DENSITY OF STARCHES AND CEREAL PRODUCTS AS A FUNCTION OF TEMPERATURE AND PRESSURE

Ch. Millauer, G. Rosa and R. Schär

BÜHLER AG, BAHNHOFSTRASSE; 9240 UZWIL, SWITZERLAND

ABSTRACT

It has been shown that milled cereal grain products have a density behaviour very similar to that of plastics. For this reason, milled grain products are often called bio-polymers. Variations in input energy and different extruding conditions (expanded, non-expanded) produced only slight density deviations. Furthermore, comparative values for various grain varieties were within a 10% range.

The linear function found for the density plotted against the temperature, pressure and moisture shows a high correlation coefficient and a very low standard error of deviation. This function was therefore used to calculate the flow velocities in dies by entering the relevant values.

1 INTRODUCTION

To calculate the shear velocity in the extruder die, the volume flow rate of the mass is required for the particular ambient conditions, that is, the temperature and the pressure. This flow rate is calculated from:

$$\dot{V} = \frac{\dot{m}}{\rho}$$ (1)

where
$\dot{V}$: volume flow rate (cm³/min)
$\dot{m}$: mass flow rate (g/min)
ρ : density (g/cm³)

It is therefore obvious that the density of the mass is inversely proportional to the volume flow rate, which in turn is directly proportional to the flow rate in the die.

Thus, it is essential to know or determine precisely the density of the particular material under the processing conditions prevailing in the die in order to limit the error in calculating the flow velocity and therefore the shear stress.

At the outset, it was assumed that such values are available from the literature, in particular since several authors have already reported on viscosity measurements in the extruder die. Unfortunately, it was found that the values cited in the literature vary between 1.07 g/cm^3 and 1.50 g/cm^3 and that they have never been stated as a function of the temperature; on the other hand, only one author[1] indicated them as a function of the pressure.

2 STATE OF THE ART

The literature on starches and milled cereal grain products offers extremely sparse information on the subject, and besides, it is normally incorporated in studies on the rheology of such products.

Thus, Harmann and Harper[2] compressed maize meal by means of a hydraulic press in a compression cell to a density of 1.24 g/cm^3.

Morgan et al.[3] determined the density of defatted soya flour with a water content of 9% at 26 bar and 177°C and stated it as being 1.15 g/cm^3.

Also Jao et al.[4] examined soya dough, though with water contents of 22, 25 and 32%, basing his studies on a density of 1.33 g/cm^3, which was measured at 21 bar and which did not rise any further when higher pressures were applied. Choi et al.[5] found 1.33 g/cm^3 for rice flour at 115°C. Bhattacharya[6] used 1.20 g/cm^3 and stated it to be constant on temperature and pressure. Smith[7] used 1.30 g/cm^3 for 55% H_2O and 1.25 g/cm^3 for 45% H_2O content at 130°C. Colonna et al.[8], Yacu[9], Canovas et al.[10], and Vergnes and Villemaire[11] assumed 1.40 to 1.50 g/cm^3 while Lancaster[12] measured 1.07 to 1.20 g/cm^3 depending on water content.

Probably the most exact examination of various maize starches was carried out by Huber[1]. His examination showed the density to be a linear function of the pressure. The test was conducted in a pill press, which was heated to the desired temperature in an oven. Huber determined a regression equation with R = 0.918 and proved the density to be dependent upon the temperature, water content and starch type.

In comparison, processes for determining the density of plastics are well-known in the literature. This is because such information is required for calculations in plastics processing (e.g. injection moulding), where pressures of up to 2000 bar and temperatures up to 350°C must be applied. As a result, the equipment described could also serve as a basis for density testing of bio-polymers.

However, the pressure ranges of commercially available devices are too wide, and they are not graduated to a sufficiently fine degree to satisfy the requirements of cooking extrusion, where the maximum temperature involved is 180°C and the maximum pressure is 200 bar. The resulting error would be far too great.

3 PROCEDURE

Since it was not possible to purchase a suitable density measuring device on the market which would meet the requirements, and because suppliers were not prepared to make the necessary adaptations, it was necessary to design, construct, calibrate and put into operation our own test apparatus.

Requirements

The test apparatus had to be designed to operate across a pressure range of 10 to 200 bar and a temperature range of 40 to 200°C.

It should be possible to set and measure the temperature of the sample to an accuracy of ± 0.5°C.

The pressure in the sample should be adjustable to a precision of ± 1 bar.

The specific gravity established for the bio-polymers should be determinable to an accuracy of ± 0.01 g/cm³, which would correspond to an error of about ± 1%.

Design of Test Apparatus

Thanks to the simplicity of the method, featuring two plungers and cylinders with floating supports, it was possible to base the design of the device on that of the company SWO Polymertechnik. In addition, the sample can be removed very easily with this method simply by pressing down the cylinder body (Figure 1).

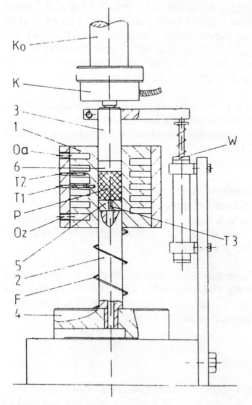

P	Sample
K	Load cell
T1}	bores for
T2}	temperature
T3}	sensors
W	Stroke measuring sensor
Oa	Oil drain
Oz	Oil feed
Ko	Piston rod of the hydraulic cylinder
F	Spring
1	Cell
2	Fixed plunger
3	Movable plunger
4	Stand
5	Bottom sealing plate
6	Top sealing plate

Figure 1   Test apparatus for p-v-T measurement

The required plunger pressure is generated by a hydraulic system using a hydraulic cylinder and a hand pump. The hand pump serves for the rough setting of the compression force. Fine adjustment and pressure compensation for the expansion of the sample volume are effected by a pressure calibration unit allowing very fine adjustment (Dreyer, Rosenkranz and Droob AG). The two plungers (2,3) each act upon the cylinder body (5, 6), which has a closer fit than the plungers in order to reduce the frictional forces and to act as a seal towards the sample (P). The measuring cell (1) was provided with a spiral-shaped ring channel in order to achieve accurate temperature control of this cell by an oil heating unit (Huber T301). The cell temperature is determined by the two sensors (T1, T2, Phillips TCA 20-10-D). The material temperature is measured by a similar sensor (T3), which is however introduced directly into the sample from below.

Figure 2   View of the test apparatus

Figure 3   Overall view of the test apparatus

Since the measuring cell and the cylinder must move in opposite directions and relative friction should therefore be minimized, the material selected for the cell was CuCoBe bronze according to DIN 1751 (Allpert 52). This material is both corrosion-resistant and has good heat conduction properties. The other components were made of X2CRNiMo 18 14 3 (1.4435). An additional advantage offered by this material combination lies in the fact that the two materials have almost the same thermal expansion coefficients (α = 0.0000176 and 0.0000175 1/k, respectively) (Figure 2 and 3).

The fittings were made with such precision that almost no bio-polymers could escape through the gap (cell, sealing plunger H4/g4; plunger H4/f6).

Performance of Test, Calculation

The pressure in the inside chamber of the cell was calculated indirectly by determining the force applied to the plunger. For this purpose, not the hydraulic pressure (see calibration), but a load cell (Burster type 8524 0-10 kN) was used, which was directly attached to the hydraulic cylinder. On the basis of the force thus determined and the cross-section of the plunger (A = 490.9 mm²), the pressure inside the cell could be calculated.

The volume of the sample filled into the chamber was determined from the plunger cross-section or inside diameter of the cell corrected by the thermal expansion value, and the plunger stroke. The plunger stroke was electronically measured by means of a stroke sensor (Greenpol LP-200 F-1). On the basis of the sample volume thus measured under processing conditions, and the input and output weights of the sample (balance: Mettler PC 4400), the density could be calculated.

All electronically measurable data were automatically read into a PC during each measurement, using appropriate measuring amplifiers and measuring boards and a data acquisition program (Labtech notebook). The PC then carried out the appropriate conversions. During the entire measuring time, all important data were displayed on-line on the screen. It was then saved and evaluated by means of a spreadsheet program.

4   MEASURING METHOD

Reference Measurement

In order to prove that the measuring device meets the standards of the state of the art in the plastics industry and yields similar results, a reference sample of PA6 (polyamide) with a cylindrical shape, a diameter of 25 mm and a height of 10 mm was produced in it; it was heated isobarically at 200 bar and then cooled again (Figure 4).

This figure shows the heating and cooling curves (see section below). In addition, it includes the values for polyamide 6 (Durethan B of the Bayer AG company) taken from the p-v-T measurement of the company SWO-Polymertechnik. These values, taken from the literature, lie exactly between those of heating and cooling, indicating that the curves were averaged.

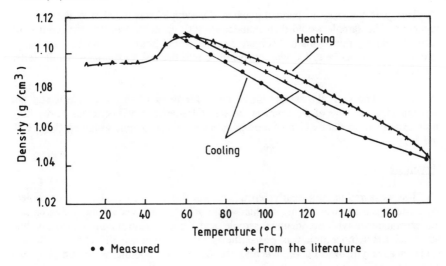

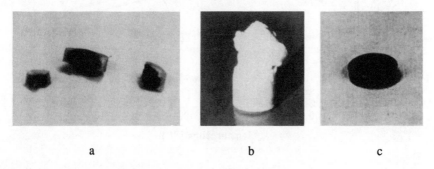

Figure 4 Reference measurement and comparison with data from the literature, polyamide 6

According to Bayer AG, it is normal to use the hysteresis cooling curve. If it was in fact used, the maximum deviation from cooling at 160°C is given by:

$$F = (1.058 - 1.066) / 1.058 \cdot 100\% = -0.75\% \qquad (2)$$

Thus, the maximum possible error relative to our measuring device is 0.75%, a value well within the range of accuracy of ±1% specified for the density.

Influence of Air Drying

Since the water content of the samples was 17 to 35%, there is of course the danger of water escaping at the high temperatures applied.

Figure 5 Handling of the sample
a: sample filled into hot test apparatus at 110°C
b: sample removed at 150°C (expanded) from test apparatus
c: sample filled into and removed from cold test apparatus

This was the case in particular when the sample was filled into the hot test apparatus. The sample would then immediately dry up at the contact surface and be unable to plastify during the test (Figure 5a), but would burn up in the middle. Therefore, the material to be tested was always filled into the test apparatus while the latter was still cold.

In the same way, the sample must not be removed when the device is still hot otherwise the pressure of the water in the starch framework will be released, resulting in an expanded product which cannot be subjected to an output weighing (Figure 5b and 5c).

Hysteresis

The measured density clearly depends on whether the sample is heated or cooled. With heating, the density is higher than with cooling. This measurement was carried out several times with the identical sample, and there was no doubt whatsoever that whenever the temperature was raised the density approached the previous density value in an asymptotic curve, whereas the curves for cooling were superimposed across the entire range.

The resulting maximum relative deviation in the hysteresis values measured during heating at, for example, 100°C was:

rel. deviation in hysteresis values = $0.0166/1.32 \cdot 100\% = 1.26\% \pm 0.63\%$ (3)

The distance of the hysteresis curve did not show any systematic influence either relative to the pressure or to the water content (Figure 6/Table 1).

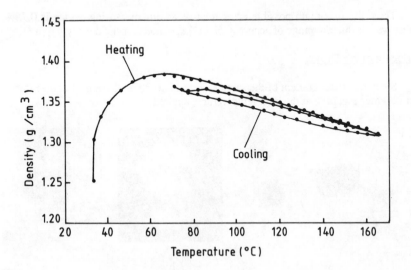

Figure 6   Hysteresis in density during heating and cooling of wheat starch with 22.5% water content at 100 bar pressure

Table 1   Hysteresis in dependence on water content and pressure

Input weight (g)	Water content (%)	Pressure (bar)				
		50	75	100	150	200
		Hysteresis (g/cm³)				
Wheat starch 5	17.5		0.016	0.016	0.014	0.017
	22.5	0.018	0.019	0.018	0.018	0.010
	24.0			0.016	0.011	

According to application engineers, a similar hysteresis is also found in the field of plastics. Their origin or cause is as yet unknown. Since only the shrinking of the material in the mould during cooling is normally of interest in plastics engineering, only the curve for the density during cooling is recorded.

Repeatability

In order to determine the repeatability of the measurements conducted, wheat starch with a moisture content of 22.5% and an input weight of 5 g and 10 g underwent the test series at a pressure of 100 bar twice and four times, respectively. The purpose was to prove that external influences of any kind do not play any part, and to check by calculation what degree of repeatability can be attained with this measuring device.

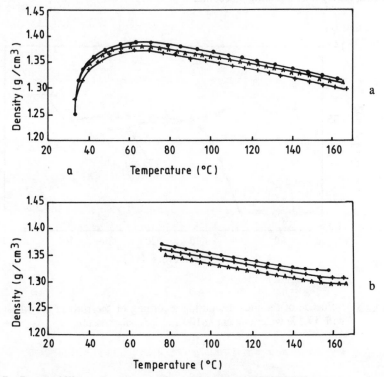

Figure 7   Repeatability measurements for wheat starch with 22.5% water content at 100 bar pressure; 5 g input weight; heating (a) and cooling (b)

Figures 7a and 7b show that with the input weight of 5 g a maximum variance of 0.02 g/cm³ occurred. The repeatability of the density values to an average density of approximately 1.35 g/cm³ was:

$$\text{Repeatability} \quad : \frac{0.02}{1.35} \cdot 100\% = 1.50\% \pm 0.75\% \tag{4}$$

Gelatinization

Since the measurement curves were recorded each time on the basis of new samples, it became obvious that the way in which the sample is gelatinized, i.e. the transition from the loose material into a plasticized mass, is decisive for the absolute density value. It was found that depending on the pressure setting, the isobars are parallel up to 0.1 g/cm³. In order to verify this fact, an isobar test was conducted, during which the pressure was increased from 100 to 200 bar and then reduced again (Figure 8).

This curve shows that when the pressure was increased by 100 bar the density rose by about only 0.01 g/cm³ in both the heating and cooling curves. On the other hand, it increased by approximately 0.07 g/cm³ in the case of the samples gelatinized and tested at 100 and 200 bar.

This means that the influence of the pressure on the density during gelatinization (type of gelatinization is decisive) was seven times greater than that of the pressure increase during the measuring procedure!

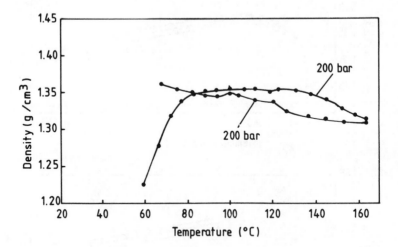

Figure 8   Influence of the pressure during recording of the isobars for wheat starch with 17.5% water content at 100 and 200 bar pressure

5   TEST RESULTS, ANALYSIS

Isobars

   Isobars of Native Wheat Starch.   Isobars for pressures of 50, 75, 100, 150 and
200 bar were recorded for native wheat starch with water contents of 17.5, 22.5 and
24%. This was done both with heating and cooling (Figures 9, 10).

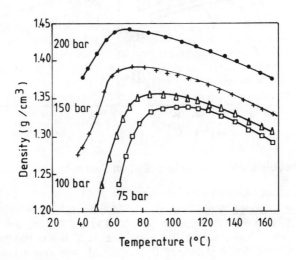

Figure 9   Isobaric heating of wheat starch with a water content of 17.5%

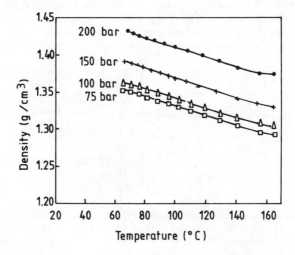

Figure 10   Isobaric cooling of wheat starch with a water content of 17.5%

Wheat Flour Isobars.   The isobaric values for 100, 150 and 200 bar were recorded for native wheat flour with water conents of of 13.0, 16.9 and 19.7%. This was done both for isobaric heating and isobaric cooling (Figure 11).

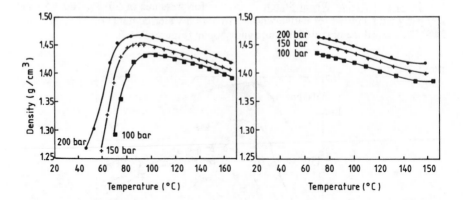

<u>Figure 11</u>   Isobaric heating and cooling of wheat flour with a water content of 13.0%

Isobars of Extruded Wheat Starch.   With extruded wheat starch with water contents of 7.9, 12.0 and 17.0% isobars for 100, 150 and 200 bar were recorded. At 7.9% water content, it was not possible to measure at 50 nor at 100 bar, because the sample was not completely plasticized (Figure 12). It was also not possible to test samples with water contents higher than 17%, because the extruded wheat starch has a very low viscosity even at low temperatures. As a consequence, it flows through the gap (20 microns) between the plunger and the cell resulting in a reduction of the sample volume and higher frictional forces, which lead to measuring errors. An attempt was made to pack the sample into foil and to place sealing covers of teflon tape or paper over and under the sample. In all cases, however, these seals did not withstand the high pressures, and the sample material again flowed into the gap.

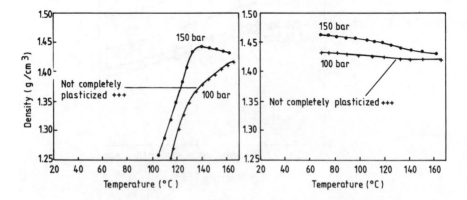

<u>Figure 12</u> Isobaric heating (a) and cooling (b) of extruded wheat starch with a water content of 7.9%

Accurately cut seals of polyamide, which were placed over and under the sample, were also unable to seal off the material, since the wheat starch became plasticized long before the plastic. As a result, the wheat starch again flowed unhindered into the gap, therefore a special sealant had to be developed.

Isotherms

The tests conducted up to this point were carried out with isobars. In order to prove the repeatability of results within a measuring cycle, extruded wheat starch samples were tested twice at isotherms of 102, 121, 152 and 182°C (time requirement approx. 2 hours!). The difference in density both of the hysteresis (increase and reduction of the temperature) and of the repetitions of a given sample were found (Figure 13) to be within the required measuring accuracy of ± 1.0% after about 2 hours.

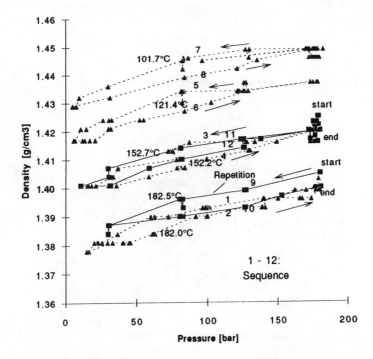

Figure 13   Isotherms of extruded wheat starch with a water content of 13.7%

Isobars and Isotherms Combined

Since all isobars recorded were always based on new samples, and since both the heating and cooling curves had to be determined, both the isobars and the isotherms were recorded in a test series with a sample to establish the density independently of heating and cooling. The isotherms proved that the density of the sample does not change under the influence of pressure and temperature during the test cycle if all the measuring is completely tight. The temperature and pressure profiles as a funtion of time are shown in Figure 14. The resulting densities were then recorded in a three-di-

mensional graph (Figure 15). On this basis of these 126 points, a regression analysis
was carried out with the SYSTAT statistics program. The results are given in Section 6.

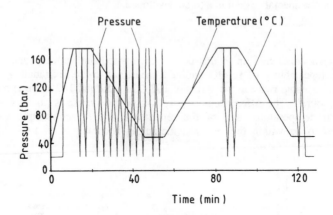

Figure 14   Pressure and temperature profiles as a function of time

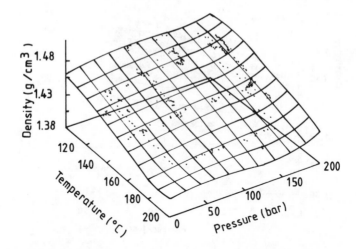

Figure 15 Extruded wheat starch, moisture content 13.7%; isobaric and isothermal
densities

Extruded Wheat Flour.   The influence of SME and water content on the resulting density was examined with the extruded wheat flour.

Specific Mechanical Energy Input.   Samples of a given raw material that were
extruded with different specific mechanical energy inputs (SME) did not produce large
differences in density. To have a clearer statement about the quantitative influence,
more examinations need to be carried out (Table 2).

<u>Table 2</u>   Density of extrudates in dependence on SME

<u>SME</u> (kWh/t)	<u>Water content</u> (%)	<u>Density</u> (g/cm³ at 150°C, 30 bar)
346	8.6	1.43
125	9.5	1.39
125	8.6	1.40 (adjusted)

<u>Water content.</u>   Providing that starch and water do not interact in the form of swelling or dissolving, the density must behave according to the law of mixtures (equation 5).

Density
$$\rho = \frac{m_{H_2O} + m_{DM}}{V_{DM} + V_{H_2O}}$$
(5)

where:

m_{H_2O} = mass of the water

m_{DM} = mass of the dry matter (g)

V_{H_2O} = volume of the water (cm³)

V_{DM} = volume of the dry matter (cm³)

ρ = density of the product (g/cm³)

Water content
$$\frac{a}{100} = \frac{m_{H_2O}}{m_{DM} + m_{H_2O}}$$
(6)

where:
a = water content (%)

Density from law of mixtures and water content

$$\rho = \frac{\rho_{H_2O} \cdot \rho_{DM}}{\frac{a}{100}(\rho_{DM} - \rho_{H_2O}) + \rho_{H_2O}}$$
(7)

The densities of wheat flour extrudates based on different water (moisture) contents and the law of mixtures are shown in Figure 16. The calculated values correspond very well with the measured values.

The density obviously depends to a large extent on the water content, as is reflected by the law of mixtures.

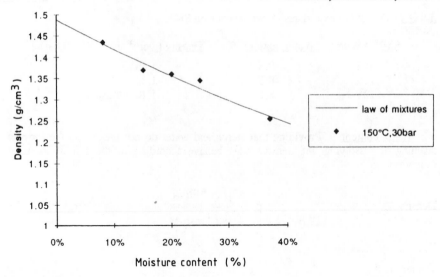

<u>Figure 16</u> Wheat flour extrudates with different water contents

<u>Extruded Maize, Degree of Gelatinization</u>.   Within the measuring accuracy, the density is not influenced by the prior history of the sample, which is reflected in the different degrees of gelatinization (Table 3).

<u>Table 3</u> Densities of maize with different degrees of expansion

Maize	Water content after		SME	Density after conditioning	Density adjusted to
	extrusion	conditioning		150°C, 30 bar	11.9% H$_2$O
	%	%	kWh/t	g/cm^3	g/cm^3
Gelatinized, slightly expanded	23.4	11.9	203.5	1.378	1.38
Expanded	8.9	14.3	298.8	1.386	1.40

<u>Other Cereal Grains, Influence of Raw Material</u>.   The density of different raw materials was determined at the same pressure, temperature and water content. Wherever the values of the same influencing factors were not available, these were determined on the basis of the regression equations. Figure 17 shows that the extruded raw materials all have different densities. The density of native starch, is clearly lower than that of the extrudate. In other words, the density is affected by the intensive shearing action that takes place during the extrusion process.

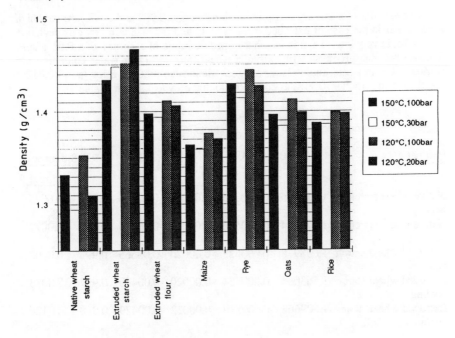

<u>Figure 17</u> Densities of different cereal grains under identical conditions

6   REGRESSION

Regression analysis were carried out by means of the SYSTAT statistics program for the measured values recorded. The following linear functions of the densities plotted against the temperature, pressure and moisture content were obtained:

$$\rho = \rho 0 + k1\vartheta + k2P + k3a \qquad (8)$$

where:
ρ = density (g/cm³)
ϑ = temperature (°C)
p = pressure (bar)
a = moisture content(%)

With the aid of this regression analysis, it was possible to calculate both the density and the coefficients k_1, k_2 and k_3. Furthermore, the correlation coefficient R, the squared correlation coefficient R^2, the total standard error of estimate and the standard error deviations of the individual coefficients were determined. The density as a function of the water content was assumed to be linear, in spite of the non-linear law of mixtures, since the deviation in the range examined is no greater than one tenth of the measuring error.

The R^2 values for this linear approximation function, 0.997 and 0.962, are very high in particular in the case of extruded wheat starch, and the standard error of deviation of 0.0075, is very low, i.e. it is within the specific range of accuracy of ± 0.01 g/cm^3. Because the measurements with extruded wheat flour at higher water contents were carried out with a seal, there resulted higher frictions so that the error grew (0.0126) and R^2 became smaller (0.918) as compared to that of wheat starch (Table 4).

Table 4   Regression analysis for various products

Product	ρ_0	k_1	k_2	k_3	R^2	Std.error of estimate
Native wheat starch heating	1.413175	-0.000765	0.00054	-0.00155	0.962	0.0075
Native wheat starch cooling	1.373254	-0.000636	0.00055	-0.00104	0.963	0.0072
Extruded wheat starch heating	1.608978	-0.000623	-0.00020	-0.00665	0.993	0.0002
Extruded wheat starch cooling	1.563066	-0.000554	-0.00020	-0.00430	0.997	0.0013
Extruded wheat flour heating and cooling	1.509663	-0.000526	0.00012	-0.00429	0.918	0.0126

The coefficient k_1, which represents the thermal expansion, is similar for almost all of the cases, with the value for heating always being higher than that for cooling. The sign of this coefficient is always negative, i.e. the volume increases and therefore the density declines as the temperature rises.

The influence of the pressure on density is found to be small as the coefficient k_2 shows, especially in case of extruded material. This is probably due to the fact that the material is easily gelatinized. This reflects that the history of the material is decisive for this density measurement.

For of all the materials tested, the water content had the greatest influence on density. The coefficient k_3 was much higher than the coefficients k_1 and k_2 respectively.

7   REFERENCES

1. D. Huber, 'Untersuchungen über die Kochextrusion von Stärken', Dissertation, ETH Zürich, 1985, p. 48.

2. D.V. Harmann and J.M. Harper, Journal of Food Science, 1974, 39, 1099.

3. Morgan et al., Journal of Food Process Engineering, 1978, 1, 74.

4. Jao et al., Journal of Food Process Engineering, 1978, 2, 101.

5. Y. Choi and M.R. Okos, 'Thermal Properties of Liquid Foods: Physical and Chemical Properties of Food', American Society of Agricultural Engineers, St. Joseph MI, 1986, p. 35.

6. M. Bhattacharya and M. Padmanabhan, 'On-line Rheological Measurements of Food Dough During Extrusion Food', (Eds. J.L. Kokini, Chi.-T. Ho and M.V. Karwe), Food Extr. Science and Techn., M. Dekker, New York, 1992, p. 224.

7. A.C. Smith, 'Studies on the Physical Structure of Starch-Based Materials in the Extrusion Cooking Process', (Eds. J.L. Kokini, Chi.-T. Ho and M.V. Karwe), Food Extr. Science and Techn., M. Dekker, New York, 1992, p. 584.

8. P. Colonna, J.L. Doublier, J.P. Melcion, F. De Monredon and C. Mercier, 'Physical and Functional Properties of Wheat Starch after Extrusion-cooking and Drum-drying: Thermal Processing and Quality of Foods', (Eds. P. Zeuthen et al.), Elsevier Applied Science Publ., London, 1984, p. 96.

9. W.A. Yacu, 'Modelling a Twin Screw Co-rotating Extruder: Thermal Processing and Quality of Foods', (Eds. P. Zeuthen et al.), Elsevier Applied Science Publ., London, 1985, p. 62.

10. G.V.B. Canovas, J. Malave-Lopez. and M.J. Peleg, Food Process Eng., 1987, 10, 1.

11. B. Vergnes and J.P. Villemaire, Rheol. Acta, 1987, 26, 570.

12. E.B. Lancaster, 'Specific Volume and Flow of Corn Grits under Pressure', American Inst. of Chem. Engineering Symp., Series 67, (108), p. 30.

FACTORS AFFECTING THE WALL SLIP BEHAVIOUR OF MODEL WHEAT FLOUR DOUGHS IN SLIT DIE RHEOMETRY

J.A. Menjivar, B. van Lengerich, C.N. Chang and D. Thorniley

NABISCO FOODS GROUP, RMS TECHNOLOGY CENTER,
200 DEFOREST AVE., GAST HANOVER, NJ 07936, USA
FAIR LAWN DEVELOPMENT CENTER, 21-11 ROUTE 208, FAIR LAWN,
NJ 07410, USA

ABSTRACT

Wall slippage is one of the major sources of error in viscometric measurements of composite, multiphase fluids. Its effect on viscometric measurements of dough systems has not been thoroughly studied. The effect of wall slippage on the slit die rheometry of model wheat flour doughs has been investigated in this study. The shear viscosity function of doughs was determined in three different ways: 1) using slit dies coupled to a ZSK-30 Werner-Pfleiderer twin screw extruder, 2) adapting a slit die to be used in conjunction with a Model 4202 Instron Universal Testing instrument, and 3) a Rheometrics Mechanical Spectrometer (RMS-800) set up with serrated parallel plate fixtures. Shear viscosity results obtained with the RMS-800 are used as the reference values for comparison with results from the slit dies. The effects of solvent level, slit die height and width, method of dough mixing, and enzymic treatment were also evaluated. Results of this study suggest that wall slip occur in slit die measurements of these model dough systems. Wall slip increases with decreasing slit height and aqueous solvent level. Enzymic treatment with protease minimizes wall slip, while treatment with hemicellulase does not reduce slippage, but increases the difference between shear viscosity measurements conducted in slit dies and parallel plate rheometers. This behaviour is explained on the basis of differences in elastomeric character of the doughs when treated with these two different enzymes. No significant differences in slip behaviour or shear viscosity profiles were found between continuously mixed (ZSK-30) or batch mixed doughs (Pin Mixer).

1 INTRODUCTION

Cereal chemists have long been attempting to use viscometric measurements to study the rheological behaviour of wheat flour doughs[1-3], with the idea of improving upon established empirical physical testing methods. This effort has not been completely successful for various reasons, not all of them will be discussed in this paper. One of these reasons is the limited experience using viscometric measurements to characterize the rheological behaviour of wheat flour doughs. Wheat flour doughs represent a complex composite and multiphase system[4] which can give rise to several problems during viscometric measurements. Literature from other fields indicate that multiphase systems do often exhibit slippage behaviour at the interface between the sample material and the testing cell[5]. Suspensions also exhibit a behaviour that is commonly termed the "sigma effect", which involves the migration of particulates away from the interface into the bulk of the flowing fluid, giving rise to a layer of lower viscosity fluid at the interface[6].

Identifying and understanding these non-idealities in viscometric flows of wheat flour doughs is critical to the performance of successful fundamental rheological measurements, this type of basic work however has not been undertaken for wheat flour doughs.

This study deals with the so called "non-slip" assumption in viscometric rheological measurements. One key prerequisite to obtain meaningful rheological results, is that the sample fluid under consideration adheres to the surface of the measuring cell, such that at the interface, the velocity of the fluid is essentially the same as that of the testing cell surface at the interface. When this requisite is not satisfied, rheologists refer to it as "slippage" or "wall slip"[7]. This phenomenon is schematically illustrated in Figure 1, where the slip velocity or fluid velocity at the boundary is termed Us, Figure 1 also illustrates that in the simplest of cases the total volumetric flow rate (Q) is made up of the contributions from the slip layer (Qs) and the bulk of the fluid (Qd). It also shows the classical Mooney equations to calculate slip velocity and the contribution of slip flow to the total fluid flow, and finally the Mooney slip correction for shear viscosity[8]. Slippage does give rise to lower shear rates than expected, and as a result, lower levels of shear stress than expected at the nominal shear rate, and therefore lower .viscosity. This is graphically illustrated in Figure 2. In the great majority of rheological studies, investigators assume that the non-slip assumption is met; in this study however, this assumption is challenged by comparing slit die results of model dough systems to measurements with a serrated parallel plate system which minimizes wall slip. Also, slit dies of different heights are used in order to test the non-slip assumption. Moreover, other factors that may affect the wall slip behaviour of model doughs are studied, such as: dough water content and the action of two hydrolytic enzymes, a protease and a hemicellulase.

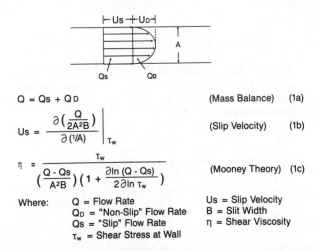

$$Q = Q_S + Q_D \qquad \text{(Mass Balance)} \quad (1a)$$

$$U_S = \left. \frac{\partial \left(\frac{Q}{2A^2B} \right)}{\partial \left(1/A \right)} \right|_{\tau_w} \qquad \text{(Slip Velocity)} \quad (1b)$$

$$\eta = \frac{\tau_w}{\left(\frac{Q - Q_S}{A^2B} \right) \left(1 + \frac{\partial \ln \left(Q - Q_S \right)}{2 \partial \ln \tau_w} \right)} \qquad \text{(Mooney Theory)} \quad (1c)$$

Where:
Q = Flow Rate U_S = Slip Velocity
Q_D = "Non-Slip" Flow Rate B = Slit Width
Q_S = "Slip" Flow Rate η = Shear Viscosity
τ_w = Shear Stress at Wall

<u>Figure 1</u>   Schematics of wall slip in slit die including Mooney[8] equations to calculate slip velocity (Us) and to correct shear viscosity for slippage at the interface

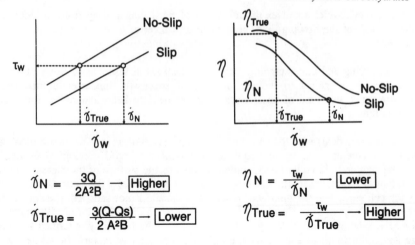

Typical experimental manifestations of wall slip in flow profiles. a) Shear stress vs. shear rate. b) Shear viscosity vs. shear rate

2   MATERIALS AND METHODS

Soft wheat flour was used to prepare model doughs which were made with an aqueous solution of 2.5 wt% sodium chloride at three different flour/solvent ratios: 100/50, 100/55 and 100/60 on a weight basis (Figure 3). Doughs were prepared by two different methods: a Werner-Pfleiderer ZSK-30 extruder and a National Manufacturing Co. 500 g pin mixer. Measurements were made in three different systems as follows (Figure 4):

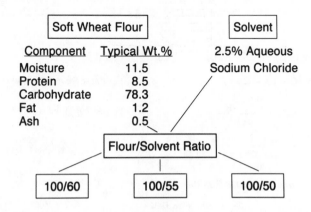

Figure 3   Model dough systems used in this study

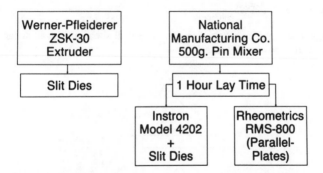

<u>Figure 4</u>   Methods for dough preparation and shear viscosity measurements

Doughs Prepared and Driven by ZSK-30 Werner-Pfleiderer Extruder

A slit die was attached to a ZSK-30 Werner-Pfleiderer co-rotating twin screw extruder for the study where doughs were made by continuous mixing. The extruder drove the dough through the slit die attached at the end of the extruder. The two sets of slit dies had the same length (L = 250 mm) and different widths: B = 40 mm and 60 mm, respectively. Two different slit heights (A = 3 mm and 4 mm) were examined for the first set of 40 mm width dies, and three different heights (A = 2 mm, 3 mm and 4 mm) were studied for the second set of 60 mm dies. The mass flow rate in the extruder was varied from 10 lb/h to 60 lb/h in order to develop different shear rates in the die. In order to maintain a constant fill of the tested material in the extruder, the ratio of screw speed to mass flow rate was kept constant at a ratio proportional to 45 RPM screw speed to 10 lb/h mass flow rate. Dough temperature was maintained constant at 25°C by adjusting the extruder barrel temperature as necessary. Five pressure transducers (Dynisco Model PT-415) were mounted on the slit die at the following locations: 15 mm, 70 mm, 125 mm, 180 mm and 235 mm measured from the die entrance. Once the pressure profile had equilibrated, pressures were recorded every minute in a data logger.

Dough Driven by Instron Model 4202 Universal Testing Machine

Model doughs prepared for measurements with the Instron were prepared using a 500 g pin mixer (National Manufacturing Co.) and mixed for 1.5 minutes at ambient temperature. All doughs were held at room temperature for one hour before measurement. For this study, a 35 mm diameter reservoir was coupled to slit dies of a constant length (L = 250 mm), constant width (B = 40 mm) and three different heights (A = 1.5 mm, 3.0 mm and 4.0 mm).

Rotational Rheometer Measurements

For this study, doughs were prepared the same way as mentioned above. A Rheometrics (Rheometrics, Inc, Piscataway, NJ) Mechanical Spectrometer (RMS-800) was used to determine shear viscosity - shear rate profiles for model doughs. A set of 25 mm diameter serrated plates were used for all measurements with a gap height of 2 mm. Shear rates studied ranged from 0.01 /s to 9.467 /s.

Computation of Flow Curves

Slit Dies The expressions used to calculate wall shear rate, wall shear stress, and shear viscosity are shown in Figure 5.

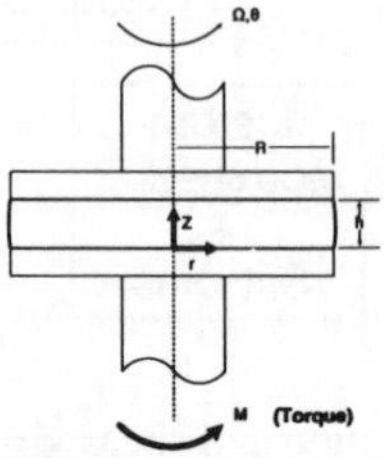

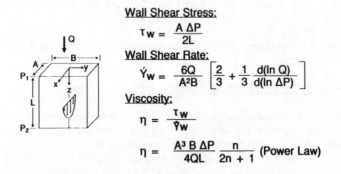

Shear Rate (at Perimeter):

$$\dot{Y}_R = \frac{\Omega R}{h}$$

Shear Stress:

$$T_{12} = T_{\theta z} = \frac{M}{2\pi R^3}\left[3 + \frac{d(\ln M)}{d(\ln \dot{Y}_R)}\right]$$

Figure 5   Schematics and governing equations for parallel plate rheometer[9]

Serrated Parallel Plates: The expressions used to calculate shear rate and shear stress are shown in Figure 6.

Wall Shear Stress:

$$\tau_w = \frac{A\,\Delta P}{2L}$$

Wall Shear Rate:

$$\dot{Y}_w = \frac{6Q}{A^2 B}\left[\frac{2}{3} + \frac{1}{3}\frac{d(\ln Q)}{d(\ln \Delta P)}\right]$$

Viscosity:

$$\eta = \frac{\tau_w}{\dot{Y}_w}$$

$$\eta = \frac{A^3 B \Delta P}{4QL}\frac{n}{2n+1} \quad \text{(Power Law)}$$

Figure 6   Schematics and governing equations for slit die rheometer[9]

Thorough discussions on the fluid mechanics theory of these flows can be found in recent reviews[9]. In this study, the derivative term in the equation for shear rate in Figure 5, and in the equation for shear stress in Figure 6 were used as in the limiting case for a Newtonian fluid.

3   RESULTS AND DISCUSSION

<u>ZSK-30 Extruder/Slit Die Studies</u>

    <u>Slit Width = 40 mm.</u>   Figure 7 illustrates a pressure profile representative of those obtained for the three model systems with the 40 mm set of slit dies. As illustrated in Figure 7, the pressure profiles are linear, indicative of fully developed flow. One of the requisites for viscometric measurements is to obtain a fully developed, steady state flow, for which a minimum length to height ratio is required, generally larger than 50. For this study, L/A ranged between 63 and 167, which as shown by Figure 7 proved sufficient for fully developed flow in the slit. Figure 8 shows the effect of flour to solvent ratio (F/S) on the flow curves of the dough systems. As expected, the lower the moisture content of the dough the higher the shear stress - shear rate curve.

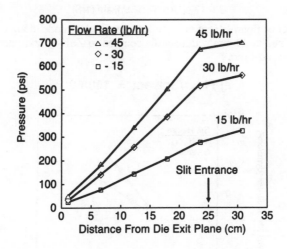

<u>Figure 7</u>   Representative pressure profile obtained in 40 mm width slit die coupled to ZSK-30 extruder. Flour/solvent ratio = 100/55, slit height = 3 mm

    The effects of slit height and dough moisture content are shown in Figures 9a through 9c, where it can be observed that for the model system with the higher moisture content (Figure 9a), there is a significant difference between the flow curves obtained with the 3 mm and 4 mm height dies. The flow curve for the 3 mm height die is lower than flow curve for the 4 mm height die. On the other hand, for the lower moisture content model doughs (Figure 9c), good agreement is obtained between the two different slit heights. These results suggest that slippage of the dough sample occurs at the sample die interface, and that this slip behaviour is more prominent for smaller slit die heights and higher dough moisture contents. Since wall slip is a surface phenomenon, the larger surface to volume ratio of the flowing sample, the more relevant does the effect of wall slip become to the overall flow of the dough sample. The surface to volume ratio of the dough sample is inversely proportional to slit height (A), therefore, more slip should be expected at smaller slit heights. This is indeed the effect observed in Figures 9a through 9c.

Plant Polymeric Carbohydrates

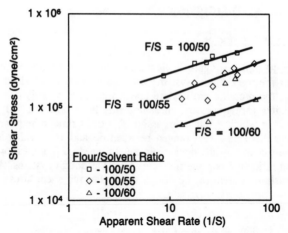

<u>Figure 8</u>   Effect of flour/solvent ratio on flow profiles of model dough systems. Data
            obtained in 40 mm width slit die coupled to ZSK-30 extruder, slit height =
            3 mm and 4 mm

a) Flour/Solvent = 100/60

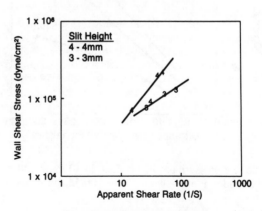

b) Flour/Solvent = 100/55

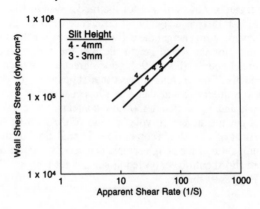

c) Flour/Solvent = 100/50

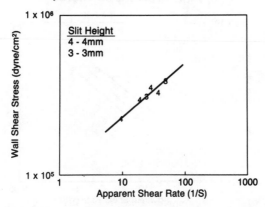

Figure 9   Effect of slit height on flow profiles of model dough systems obtained in 40 mm width slit die coupled to ZSK-30 extruder: a) F/S = 100/60, b) F/S = 100/55, c) F/S = 100/50

Slit Width: 60 mm.   Figures 10a through 10c illustrate the effect of slit height and dough moisture content on the flow behaviour of the model doughs in the 60 mm width slit dies. A similar behaviour to that found in the 40 mm width dies is observed, where wall slip seems to be present at the higher solvent contents (Figure 10a, 100/60) as evidenced by the effect of slit height on the flow curves. This effect of slit height decreases with decreasing content of water in dough, until at a flour to solvent ratio of 100/50, the flow curves obtained with the three different slit heights (2 mm, 3 mm, and 4 mm, respectively) superimpose on each other (Figure 10c). These results also indicate that slit width to height used in this study is satisfactory and apparently sufficient to develop a two dimensional and not a three dimensional flow. This is also shown in Figures 11a through 11c where the viscosity curves for the two sets of slit dies (40 mm and 60 mm width) are plotted for each of the solvent levels.

a) Flour/Solvent = 100/60

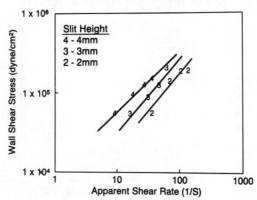

b) Flour/Solvent = 100/55

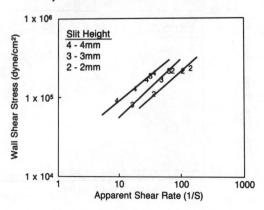

c) Flour/Solvent = 100/50

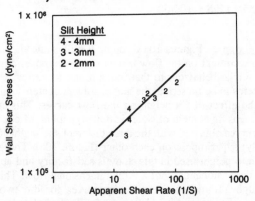

<u>Figure 10</u> Effect of slit height on flow profiles of model dough systems obtained in
60 mm width slit die coupled to ZSK-30 extruder: a) F/S = 100/60, b) F/S
= 100/55, c) F/S = 100/50

a) Flour/Solvent Ratio = 100/60

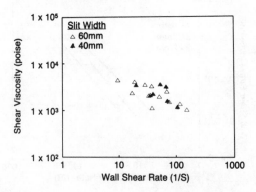

b) Flour/Solvent Ratio = 100/55

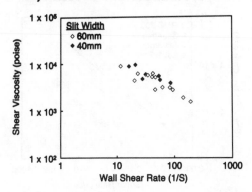

c) Flour/Solvent Ratio = 100/50

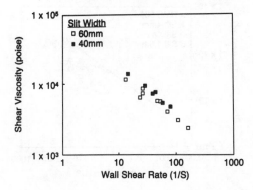

<u>Figure 11</u> Effect of slit width on shear viscosity - shear rate curves of model dough systems: a) F/S = 100/60, b) F/S = 100/55, c) F/S = 100/50

<u>Instron Model 4202 - Slit Dies Measurements: Comparison to RMS-800 Results</u>

For this study, the focus of the investigation was on the comparison between the flow curves obtained with the slit dies coupled to the Instron Model 4202 instrument, and the flow curves obtained with serrated parallel plates (RMS-800). Dough samples used for these measurements were prepared in a batch pin mixer.

Figures 12a through 12c illustrate the results obtained for the three model doughs. It can be noticed that the difference between the slit dies and the RMS-800 flow curves diminishes as flour/solvent ratio increases (decreasing level of solvent). This result is consistent with the effect of dough moisture on wall slip found in the studies conducted with the ZSK-30 extruder. In this case however, the slit dies curves never match perfectly the RMS-800 curves, and better agreement is found at the higher rates. More research is needed in order to better understand the differences between these two viscometric methods at the higher F/S ratio.

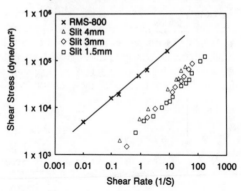

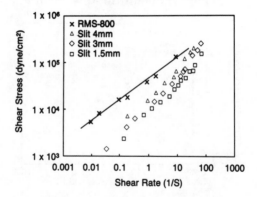

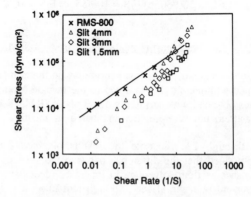

<u>Figure 12</u> Comparison between flow curves obtained in slit dies (width = 40 mm) coupled to Instron Model 4202 instrument and flow curves obtained with serrated parallel plate rheometer (RMS-800): a) F/S = 100/60, b) F/S = 100/55, c) F/S = 100/50

Overall Comparison of Slit Die Measurements with Serrated Parallel Plate Measurements

Figure 13 summarizes the shear viscosity results obtained on the F/S = 100/50 dough model system. In this plot results obtained with the ZSK-30, Instron 4202, and RMS-800 are illustrated. It can be observed that the best agreement between all these measurements is obtained at the higher rates (> 1/s). The disagreement at lower rates is partly due to limitations on the sensitivity of the pressure transducers at lower shear stresses.

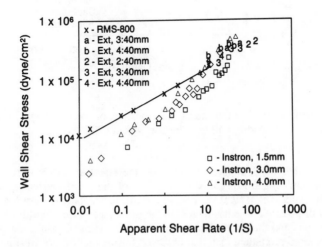

Figure 13 Comparison of flow curves obtained on F/S = 100/50 model dough system with parallel plate rheometer (RMS-800) and slit die sets coupled to ZSK-30 extruder and Instron Model 4202 instrument

Other possible reasons for the disagreement between results obtained with the slit dies and the parallel plate geometry may have to do with the rupture behaviour of this lightly elastomeric dough[10]. Parallel plate measurements on these model systems are obtained developing stress-strain curves at a constant shear rate. However, in most cases, steady state is not achieved due to the rupture of the dough sample at finite strains. Beyond rupture, sample instabilities are observed within the parallel plates gap, which prevents the attainment of steady state. Therefore, reported flow curves for the parallel plate rheometer results are obtained from the maximum stress, as opposed to the steady state wall stress. The impact of this effect on the flow curves is difficult to evaluate.

Effect of Hydrolytic Enzymes on the Slip Behaviour of Model Doughs

Figure 14 illustrates that the addition of 0.01% (based on flour weight) protease to the 100/50 dough, minimizes the difference between the RMS-800 and the slit die curves obtained with the Instron setup; also, notice that the effect of slit height seems to be minimized.

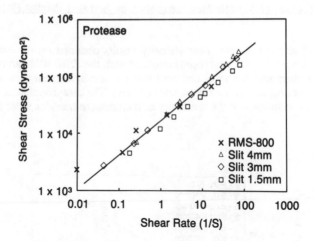

<u>Figure 14</u> Effect of 0.01% protease on flow profiles of model dough system (F/S = 100/50, slit width = 40 mm, Instron Model 4202 set up)

On the other hand, Figure 15 illustrates the effect of 0.023% (based on flour weight) of a hemicellulase on the slip behaviour of the model system containing the least formula water, i.e. 100/50 flour solvent ratio. It is apparent from Figure 15 that the addition of hemicellulase has increased the difference in flow profiles between the RMS-800 results and the slit die (Instron setup) measurements. Since the effect of slit height is not more prominent than that observed for the control dough (Figure 12c), it is not clear that the difference between the slit die and the RMS-800 results is due to wall slip effects.

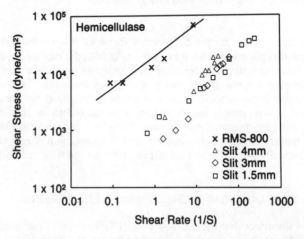

<u>Figure 15</u> Effect of 0.023% hemicellulase on flow curves of model dough system (F/S = 100/50, slit width = 40 mm, Instron Model 4202 set up)

The effect of the protease (papain, endopeptidase) on the dough system is to breakdown the protein network and decrease the rubber-like elastomeric properties of dough. On the other hand, the effect of the hemicellulase is to breakdown the pentosan network, thus enhancing the rubber-like elastomeric properties of the dough. RMS-800 measurements on the dough treated with hemicellulase exhibit a more prominent rupture behaviour than those treated with protease, which exhibit a more plastic behaviour. This difference in elastomeric character between doughs treated with protease and hemicellulase is believed to be at least partially responsible for the difference in flow curves observed in Figure 15.

4 CONCLUSIONS

The results of this study suggest that under flow, these model dough systems slip at their interface with slit dies. The slippage at the wall decreases in most cases with increasing slit heights (lower sample surface to volume ratio), and decreasing level of aqueous solvent. These results are consistent with previous findings on other multiphase suspensions, and indicate the need to pay particular attention to slip considerations when performing viscometric measurements on dough systems.

The results of this study also suggest that the slip phenomenon in dough systems might be at least partially associated with the elastomeric character of these systems[7]. Upon the breakdown of the protein network by enzymic treatment, the effect of slit height and the differences between parallel-plate and slit die measurements do decrease dramatically. Conversely, upon the enhancement of elastomeric effects of the dough systems through the use of a hemicellulase, the effect of slit height is apparent, and the difference between RMS-800 and slit die measurements is increased significantly. This increase in the difference between the measurements performed by these two techniques, is most likely due to the rupture behaviour of elastomeric doughs in the parallel-plate measurements, which precludes the attainment of steady state flow in this configuration. These differences between the two viscometric techniques warrant further investigation.

Finally, two different mixing methods were used in this study, the shear viscosity results indicate that there are only relatively small differences between the dough systems prepared by the two methods. Since the two methods can provide a very different structure to the final dough systems, it is not clear from this study whether shear viscosity is sensitive enough to possible microstructural differences between their doughs. More research is needed in this direction.

5 REFERENCES

1. G.E. Hibberd and N.S. Parker, Cereal Chem., 1975, 52, 1r-23r.

2. A.H. Bloksma and W. Bushuk, 'Rheology and Chemistry of Dough: Wheat: Chemistry and Technology', (Ed. Y. Pomeranz), Am. Assoc. Cereal Chemists, St. Paul, MN, 1989, Vol. 2, Chap. 4.

3. S.J. Dus and J.L. Kokini, J. Rheology, 1990, 34, 1069.

4. A.H. Bloksma, Cereal Foods World, 1990, 35, 237.

5. D.M. Kalyon and U. Yilmazer, 'Rheological Behaviour of Highly Filled Suspensions which Exhibit Slip at the Wall: Polymer Rheology and Processing', (Eds. A.A. Collyer and L.A. Utracki), Elsevier Applied Science, London and New York, 1990, pp. 241-275.

6. D. Leighton and A. Acrivos, J. Fluid Mech., 1987, 181, 415.

7. J.M. Dealy and K.F. Wissburn, 'Melt Rheology and Its Role in Plastics Processing', Van Nostrand Reinhold, New York, 1990, pp. 158, 305, 340.

8. M. Mooney, J. Rheol., 1931, 2, 210.

9. F.W. Macosko, 'Fluid Mechanics Measurements in Non-Newtonian Fluids: Fluid Mechanics Measurements', (Ed. R.J. Goldstein), Hemisphere Pub. Corp., New York, 1983, Ch. 9.

10. J.A. Menjivar, 'Fundamental Aspects of Dough Rheology: Dough Rheology and Baked Product Texture', (Eds. H. Faridi and J. Faubion), Van Nostrand Reinhold Publishers, New York, 1990, p. 1.

Nutrition

NUTRITIONAL IMPORTANCE AND CLASSIFICATION OF FOOD CARBOHYDRATES

N.-G. Asp

DEPARTMENT OF APPLIED NUTRITION AND FOOD CHEMISTRY, UNIVERSITY OF LUND, CHEMICAL CENTER, P.O. BOX 124, 221 00 LUND, SWEDEN

1 INTRODUCTION

Carbohydrates are the main source of energy in most human diets, constituting from less than 40 to 80% of the calories. Current dietary guidelines recommend that the fat intake in Western countries should be decreased from around 40% at present to not more than 30% of the energy intake (e.g.[1,2]). The protein should be kept at about the present level, corresponding to 10 - 15% of the energy. Thus, the carbohydrate intake has to be increased to at least 55 - 60% of the energy.

Originally, the recommendation to increase the carbohydrate intake came as a consequence of the fat and protein recommendations. In recent years, however, the specific nutritional importance of the carbohydrates as such has been more and more emphasized, and new developments call for a more specific nutritional classification of the different food carbohydrates as a basis for more specific recommendations. Labelling of foods regarding carbohydrate content is a separate, but closely related issue. The "carbohydrate by difference" figure, that is still prevailing on food packages, needs to be replaced by a number of different nutritionally relevant carbohydrate fractions.

Table 1   Main food carbohydrates

Monosaccharides	Polysaccharides
Glucose	*Starch*
Fructose	- amylopectin
	- amylose
Disaccharides	- modified food starches
Sucrose	*Inulin*
Lactose	*Non-starch polysaccharides (NSP)*
	- cellulose
Oligosaccharides	- hemicellulose
α-Galactosides	- pectins
- raffinose, stachyose	- gums
Fructans	- mucilages
- fructooligosaccharides	- algal polysaccharides

Table 1 shows the quantitatively most important food carbohydrates. Starch generally occurs in the largest amount, followed by sucrose and - when milk products are consumed - lactose.

2   CARBOHYDRATES AS A SOURCE OF ENERGY

Food carbohydrates that are digested and absorbed in the small intestine supply monosaccharides - mainly glucose - to the blood. This glucose can be utilized in three different ways:

1) oxidation in various tissues - the brain is dependent upon glucose as a source of energy, whereas other tissues can use glucose or fatty acids,
2) storage as glycogen in the liver and muscles. Liver glycogen can be reused to provide blood glucose, whereas muscle glycogen is utilized as a source of energy only in the muscle, and
3) conversion to fatty acids and storage as body fat.

Not long ago, a general opinion was that bread and potatoes are fattening because of their high carbohydrate content. However, there is now considerable evidence that dietary carbohydrates generally do not cause a net increase in an individual's fat content. Only a small part of excess dietary carbohydrates is converted to fat in the body, and this occurs in the liver rather than in the adipose tissue[3-5]. Excess fat, on the other hand, is readily stored in the adipose tissue. Thus, restriction of the fat intake - in accordance with current dietary guidelines - is the most important measure to maintain or decrease the body weight.

3   RATE OF CARBOHYDRATE DIGESTION AND ABSORPTION

Diabetic patients have since long been advised to choose carbohydrates that are slowly digested and absorbed, giving a limited and sustained blood glucose elevation with minimum insulin requirement. In addition to choice of "slow" carbohydrates, restrictions in the amount of dietary carbohydrates have been recommended as well.

Generally it has been believed that starch is slowly digested and absorbed due to its high molecular weight ("complex carbohydrate"). Sucrose and other low-molecular weight carbohydrates ("simple sugars"), on the other hand, have been regarded as rapidly absorbed. It is remarkable that this view has been so prevalent in spite of lack of scientific evidence. On the contrary, data accumulated in the seventies showing that the height and shape of the blood glucose curve could be quite different after intake of various foods, and that these differences were unrelated to the molecular size of the carbohydrates. Among the low molecular weight carbohydrates, fructose gives a very low glycaemic response, and sucrose is intermediate between glucose and fructose. Starchy foods are found in the whole range of "slow" to "rapid" (for review see e.g.[6]). A number of carbohydrate and food properties determining the glycaemic response have now been identified and will be the subject of the following presentations in this session.

There is now increasing evidence that low glycaemic reponses after meals may be beneficial in various respects (Table 2). They are in fact important in the management, and may be also in the prevention of maturity onset diabetes. Furthermore, beneficial effects on satiety and athletic performance after a meal have been demonstrated[6]. A low glycaemic response after starchy foods is also correlated with a less pronounced pH drop in dental plaques, which may be of importance for the cariogenic properties[7]. Since low glycaemic and insulinaemic responses are correlated, it is also of potential interest for prevention and treatment of diseases in which hyperinsulinaemia is regarded as an important pathogenetic factor, i.e. hypertension, obesity and cardiovascular disease. Finally, quite speculatively, hyperglycaemia could be more generally related to tissue ageing[8].

Table 2   Possible beneficial effects of low glycaemic response foods

> Diabetes
> - blood glucose control
> - complications
>
> Generally
> - insulin levels
>   (hypertension, obesity, cardiovascular disease)
> - satiety
> - physical performance
> - serotonin precursors
> - tissue protein glycosylation (ageing ?)

4   SMALL-INTESTINAL DIGESTIBILITY - A KEY DETERMINANT

Whereas carbohydrates that are digested and absorbed in the small intestine provide glucose to body tissues, undigestible carbohydrates are delivered to the large intestine and fermented to various extents. As in ruminants, the main products of this anaerobic fermentation are acetate, propionate and butyrate. Acetate and propionate are absorbed and metabolized in peripheral tissues and the liver, respectively, and their possible effects on carbohydrate and lipid metabolism are currently being investigated. Butyrate is an important source of energy for the epithelial cells of the large intestine itself, and may be important in protecting against colonic cancer (for review, see e.g.[9]).

The fundamental nutritional difference between carbohydrates due to their small-intestinal digestibility was recognized by McCance and Lawrence more than 60 years ago[10]. It formed the background for their differentiation between "available" and "unavailable" carbohydrates, based on determination of starch and digestible sugars.

5   DIETARY FIBRE

Dietary fibre was first defined as the remnants of plant cell-walls not digested in the small intestine[11]. With this definition it constitutes the non-starch polysaccharides of the plant cell-walls, but also undigestible protein, inorganic material, tannins, cutin etc. The redefinition by Trowell et al.[12] restricted the definition to polysaccharides and lignin, but enlarged it to include all undigestible polysaccharides. There were two reasons for this. First, purified polysaccharides such as pectins and gums were frequently used to study physiological effects of dietary fibre constituents, and second, cell-wall polysaccharides could not easily be differentiated analytically from undigestible polysaccharides of other origin[13].

There are still differences of opinions regarding the definition of dietary fibre, the two main alternatives being as non-starch polysaccharides of plant cell-wall origin[14] or as undigestible polysaccharides plus lignin[15].

Regardless of the definition, dietary fibre or non-starch polysaccharides include a large number of polysaccharides with quite different properties, both from the chemical and physiological points of view. Insoluble, lignified types of dietary fibre have the most prominent faecal bulking effect due to their resistance to fermentation, whereas soluble, gel-forming polysaccharides are most efficient in lowering blood cholesterol and blood glucose after a meal. The physiological effects have little relationship to the monomeric composition of the polysaccharides, and the usefulness of the soluble fibre concept will be discussed by Johnson below. In any case, the solubility is highly method dependent[13].

6   RESISTANT STARCH

Resistant starch is now generally defined as the sum of starch and products of starch degradation not absorbed in the small intestine[14,15]. It is then an undigestible polysaccharide and would be dietary fibre according to the physiologically oriented definition mentioned above[12,15]. However, it is not a cell-wall component and would therefore fall outside the more restrictive definition as cell-wall non-starch polysaccharides[11,14]. Physiologically, resistant starch is obviously more similar to non-starch polysaccharides than to digestible starch.

7   OLIGOSACCHARIDES

The International Union of Pure and Applied Chemistry (IUPAC) defines oligosaccharides as having less than 10 monomeric residues, whereas the recent British Nutrition Foundation report[16] put the limit at 20. In practice, however, oligosaccharides are defined as carbohydrates soluble or extractable in aqueous ethanol. Precipitation in 78-80% ethanol (or dialysis/ultrafiltration) is generally used to separate oligosaccharides and starch degradation products from polysaccharides in dietary fibre analysis[13].

The α-galactosides in leguminous seeds and fructans in onions, artichokes etc. are quantitatively the most important groups of naturally occurring undigestible oligosaccharides. They are not included in any of the current dietary fibre definitions, although their physiological effects are similar to those of non-starch polysaccharides in many ways. Inulin is also not determined as dietary fibre with any of the current methods in spite of a degree of polymerization (DP) of 30 or more. Arabans in sugar beet fibre is another example of a dietary fibre polysaccharide that is extremely soluble in alcohol due to its extensive branching[17].

Chemically modified food starches[18], dry heat treated starch[19] and polydextrose are degraded by amylases to alcohol soluble fragments that are partly undigestible.

8 COMPLEX CARBOHYDRATES

The term complex carbohydrates was used in 1977 by the U.S. Senate Committee on Nutrition and Human Needs (McGovern Report) without any exact definition, but meaning in practice digestible polymeric carbohydrate, i.e. starch. In Britain, it was reintroduced to mean starch and non-starch polysaccharides[16]. The term is questionable nutritionally since it includes both digestible and undigestible polysaccharides. Recommendations regarding complex carbohydrate intake (British definition) are very complicated to interpret in terms of foods even for experts, and may come into conflict with recommendations of upper limits of dietary fibre intake[2].

9 ANALYTICAL CONSIDERATIONS

For scientific purposes it is essential to characterize as far as possible the carbohydrate composition of any material used to study the physiological effects of dietary fibre or non-starch polysaccharides. For instance, oligosaccharides and resistant starch fractions, that might have physiological effects similar to those of non-starch polysaccharides need to be accounted for regardless of which definition and analysis method is used for assay of the fibre.

For labelling purposes, the situation is different. Carbohydrate fractions have to be defined "as measured by a certain method" in the same way as for other nutrients. Provisional decisions have to be taken, subject to revision whenever required by the scientific development.

Both from the scientific and the public interest points of view, the first division of carbohydrates would be into "dietary fibre" and "other carbohydrates". The two most extensively tested methods for dietary fibre analysis are the enzymic, gravimetric method approved by the AOAC and the Englyst methods for non-starch polysaccharides (for review see[13]). In products containing lignin and/or resistant starch, the AOAC method gives higher values than the Englyst methods. In practice, however, the differences are small in normal foods and usually within the analytical errors of the different methods[13].

The delimitation between oligosaccharides and polysaccharides needs further consideration. For instance, inulin is not included in dietary fibre determinations with any current method, although it is an undigestible polysaccharide.

Methods for the determination of various nutritionally relevant starch fractions are currently being developed, as well as methods predicting the glycaemic response (see Björck and Englyst et al. below).

10 CONCLUSIONS

Recent developments concerning the nutritional importance of various food carbohydrates give cause for a renewed nutritional classification and more specific recommendations regarding intake of foods rich in carbohydrates.

11 REFERENCES

1. Anon., 'Nordic dietary recommendations', Statens Livsmedelsverk, Uppsala, 1989.
2. Anon., 'Diet, nutrition, and the prevention of chronic diseases', World Health Organization, Technical Report Series 797, Geneva, 1990.
3. L. Sjöström, Acta Med. Scand., 1972, Suppl. 544, 1.
4. K.J. Acheson, J.P. Flatt and E. Jéquier, Metabolism, 1982, 31, 1234.
5. K.J. Acheson, J.P. Flatt, T. Bessard, K. Anantharaman, J.P. Flatt and E. Jéquier, Am. J. Clin. Nutr., 1988, 48, 240.
6. A.S. Truswell, Europ. J. Clin. Nutr., 1992, 46.
7. P. Lingström, J. Holm, D. Birkhed and I. Björck, Scand. J. Dent. Res., 1989, 97, 392.
8. A. Cerami, H. Vlassana and M. Brownlee, Scientific American, 1989, May, 90.
9. J.H. Cummings, Scand. J. Gastroent., 1984, 20, 88.
10. R.A. McCance and R.D. Lawrence, 'The carbohydrate content of foods', Spec. Rep. Ser. Med. Res. Coun. Lond., No. 195, HMSO, London, 1929.
11. H.G. Trowell, Am. J. Clin. Nutr., 1972, 25, 926.
12. H.C. Trowell, D.A.T. Southgate, T.M.S. Wolever, A.R. Leeds, M.A. Gassull and D.J.A. Jenkins, Lancet, 1976, i, 967.
13. N.-G. Asp, T.F. Schweizer, D.A.T. Southgate and O. Theander, 'Dietary Fibre - A Component of Food, Nutritional Function in Health and Disease', (Eds. T.F. Schweizer and C.A. Edwards), Springer-Verlag, London, 1992, p. 57.
14. J.H. Cummings and H.N. Englyst, Trends Food Sci. Techn., 1991, 2, 99.
15. N.-G. Asp and I. Björck, Trends Food Sci. Techn., 1992, 5, 111.
16. Anon., 'Complex Carbohydrates in Foods', The Report of the British Nutrition Foundation's Task Force, Chapman and Hall, London, 1990, p. 3.
17. N.-G. Asp, 'New Developments in Dietary Fiber', (Eds. J. Furda and C.J. Brine), Plenum Press, New York, 1990, p. 227.
18. I. Björck, A. Gunnarsson and K. Östergård, Starch/Stärke, 1989, 41, 128.
19. M. Siljeström, I. Björck and E. Westerlund, Starch/Stärke, 1989, 41, 95.

THE GLYCAEMIC RESPONSE AFTER STARCHY FOOD CONSUMPTION AS AFFECTED BY CHOICE OF RAW MATERIAL AND PROCESSING

I. Björck

DEPARTMENT OF APPLIED NUTRITION AND FOOD CHEMISTRY, UNIVERSITY OF LUND, P.O. BOX 124, 122 00 LUND, SWEDEN

1 INTRODUCTION

Until recently, starch was believed to be slowly digested and absorbed, thus favouring low glycaemic responses. The basis for this concept was the opinion that molecular size was an important determinant of the rate of digestion. These presumed inherent features of starch were considered mainly in diabetes, and starchy foods were classified as a homogenous group with more beneficial properties than most foods containing low-molecular-weight carbohydrates. This classification prevailed to a large extent in the Swedish guidelines to diabetics also during the 1980's.

Differences to the opinion that starch was slowly digested began to appear in the early 1960's. It then became apparent that the digestion of soluble starch was in fact so rapid that the absorption of glucose was the rate limiting step[1,2]. However, despite this apparent surplus of amylolytic enzymes, Crapo and co-workers[3] found that starch in foods varied greatly in the effect on post-prandial glycaemia. This was interpreted in terms of differences in the rate of digestion of starch when present in a food matrix. A new picture then emerged suggesting that the glycaemic response to starch was influenced by various food factors.

Today it is well established that starchy foods differ in effects on blood glucose and insulin responses after a meal. Starch in potatoes and bread usually elicit high responses, whereas rice, pasta, certain muesli and, in particular legumes, produce a low postprandial glycaemia[4-7]. It should be noted that the glucose response to some starchy foods, exceeds that with sucrose or foods containing low-molecular weight carbohydrates[7]. Consequently, a nutritional classification of carbohydrates should be based on properties in foods. Such a classification was made possible by the glycaemic index (GI) concept[4]. The GI is defined as the postprandial glucose area with the test product as a percentage of the corresponding area following ingestion of an equivalent amount of available carbohydrates (50 g) in the form of glucose or white bread. Usually the 2h areas are used for calculation. A similar approach is used for ranking based on postprandial insulin response[8].

With white bread as reference (GI 100), starchy products may range in GI from 12 in certain legumes (bengal gram dal) to over 100 in the case of e.g. corn flakes[6].

2   GLYCAEMIC RESPONSES

Nutritional Implications

Important health benefits are being related to foods that evoke low blood glucose responses after a meal[9-17]. The metabolic responses to starchy foods are particularly intriguing, as starch constitutes our major source of available carbohydrate.

A low rate of glucose delivery to the blood will lower the demand for insulin. This is an important nutritional feature as hyperinsulinemia is increasingly being discussed as a risk factor in the genesis of several diseases connected with affluence, such as hypertension, hyperlipidemia, obesity and cardiovascular disease[9,10]. Diets based on foods which promote low glycaemic and insulin responses have further been shown to improve metabolic control in patients with diabetes[11,12] or hyperlipidemia[13,14]. Other suggested advantageous effects of slow release starchy foods include a longer duration of satiety after a meal[15,16] and protection against development of type II diabetes[17]. Slowly digested and absorbed carbohydrates can thus be considered beneficial both in terms of prevention and management of these diseases.

Mechanisms to Differences in Glycaemic Response

Several food factors influence the glycaemic response. Some are related to the properties of the starch as such, others to the food structure or to the presence of certain components (see Table 1). In general, the higher the degree of order within the starch moiety or food architecture, the lower the response. However, although many factors have been identified, it is still not known which factors are of major importance, or how to best make use of this knowledge to improve the nutritional properties of starch in foods. As is evident from Table 1, not only the ingredient material but also the type or extent of food processing are important variables.

The rate of starch digestion was proposed early as an important determinant of the glycaemic response[3,18-20], and most of the factors listed above have been shown to affect the rate of in vitro amylolysis. Consequently, a high crystallinity, as in raw[21,22] or firmly retrograded starch[23-25], botanical encapsulation of the starch moiety within cell[15,26,27] and tissue structures[28,29], or the presence of antinutrients capable of reducing amylase activity[30-32], are all factors obstructing the rate of digestion.

Other factors suggested to decrease the amylase availability include interactions between starch and protein or lipids[33,34]. However, the importance in vivo of such interactions in food products remains to be elucidated. Further, a high amylose/amylopectin ratio is often discussed as a factor reducing the metabolic response to starch[35-38]. Data are, however, contradictory[39,40].

The presence of viscous fibre, on the other hand, is believed to reduce glycaemia by slowing down motility and diffusion in the upper gut[41,42]. Although the first food factor identified[43], effects of viscous fibre have mainly been evaluated following enrichment[44,45], and less is known about the impact of naturally occurring levels in common food. In fact, when correlating GI's to the soluble dietary fibre content for a variety of foods, no significant correlation appeared[46]. Possibly, the preservation of the fibre matrix may be more important than the distribution of soluble to insoluble dietary fibre per se. Hence, there are several reports showing that processes leading to a disin-

<u>Table 1</u>   Food factors decreasing the postprandial blood glucose response to starch

	<u>Decrease in</u> <u>glucose response</u>
<u>Starch structure</u>	
Extent of	
gelatinization	↘*
retrogradation (amylose)	↗
interactions (lipids or protein)	↗
amylose/amylopectin ratio	↗
<u>Food structure</u>	
Extent of	
botanical integrity	↗
physical density	↗
<u>Food composition</u>	
Content of	
phytic acid, tannins, lectins	↗
viscous dietary fibre	↗

* Arrows indicate the necessary change: (↗ increase, ↘ decrease in starch structure, food structure and food composition)

tegration of the botanical tissue will increase enzyme availability, and hence the glycaemic and insulin response to starchy foods[28,29].

3   GLYCAEMIC RESPONSES TO CEREAL AND LEGUME PRODUCTS

The present paper will address some food factors of importance for the glycaemic response to starch in cereal and legume products. The possibility to improve properties, i.e. reduce glycaemia, of in particular cereal products is also discussed.

Degree of Gelatinization

Raw starch is only slowly attacked by amylases. Any process leading to gelatinization will increase the availability for enzymic digestion. Accordingly, it is well documented that raw starch elicits lower glucose and insulin responses than completely gelatinized starch[21,22]. However, the starch in most ready-to-eat products can be expected to be more or less completely gelatinized. Also, as judged from experiments in rats[47,48], there is a marked increase in glucose response already at intermediate levels of gelatinization. Further, no significant correlation was obtained between GI and the degree of gelatinization in different cereal products[49], suggesting that other food factors may be more important in realistic food items. The impact of incomplete gelatinization is thus probably confined to raw food or to products where the hydration of starch is kept at a low level, like in e.g. cookies[4,13] or muesli[4,50].

Food Texture (Pasta)

Despite an essentially complete degree of gelatinization, reports with pasta have indicated low metabolic responses[4,5,51,52]. Data with spaghetti and linguine in our laboratory verify beneficial properties of pasta, with a low and sustained net-increment in blood glucose over time in healthy subjects[53]. In contrast, a white durum bread made from 'spaghetti ingredients' not only produced a high initial glucose peak, but also a rapid decline (1.5-2h) and a prominent hypoglycaemia in the late postprandial phase (2-3h). However, despite these important differences in the course of glycaemia, there were no differences in GI between bread and pasta when basing ranking on the 2h glucose areas. The rapid decline in blood glucose after bread resulted from an overstimulation of insulin secretion. A hypoglycaemia to rapidly digested and absorbed carbohydrates has been connected with hunger sensations in man[15,16,54]. As judged from our results[53], the reliability of GI in healthy subjects is improved by excluding the late post-prandial phase, and in the following all GI's and Insulinaemic Indices (II's) are based on 0-1.5h areas.

In Table 2 GI's of pasta products, selected to cover several technological parameters, are displayed[53,55]. A white bread made from durum wheat with added monoglycerides was used as reference product.

Table 2   Glycaemic indices (GI's) to pasta products in healthy subjects

		GI
Linguine, roll-sheated; fresh		
durum + mg*(thick,	3.3·2.2 mm)	62
durum + mg*(thin,	1.2·2.2 mm)	70
durum + egg (thin,	1.2·2.2 mm)	64
Spaghetti, extruded and dried		
durum + mg*	(Ø=1.8mm)	67
Macaroni, extruded and dried		
durum/swedish wheat	(30/70)	71

* mg = monoglycerides

All pasta had lower indices (GI=56-71) than the reference bread, and no significant differences appeared between pasta products. Consequently, ingredients, food form and processing conditions can be varied within broad limits still maintaining a low glycaemic response.

The 'rapid' behaviour of starch in the white durum bread negates the view that the properties of pasta are related to the durum per se[5]. A disintegration of the spaghetti into a coarse porridge significantly increased GI by almost 20 units, by increasing the rate of starch digestion[55]. This suggests that food texture is an important determinant. However, even after disintegration, firm pasta particles remained, and the pasta porridge still produced a lower glucose response than the porous bread. Such phenomena related to the physical texture could be explored to improve properties of cereal products.

In conclusion, the low metabolic response to pasta appears to be a general phenomenon related to the dense physical texture. The nature of the enzyme barrier at a molecular level remains to be established.

Cereal Fiber / Food Structure (Bread)

The high glucose response to white wheat bread is representative of most common bread products[56]. Bread is an important source of starch in our diet, and also a rich source of dietary fiber, vitamins and minerals. This makes it a challenge to improve the properties of starch in bread.

In a study in healthy subjects, the potential of exchanging flour for intact kernels was tested[57]. In the kernel breads, 80% of the starch was provided in the form of preheated kernels from oats, wheat, rye and barley and 20% in the form of white wheat flour. A barley bread with 80% whole meal barley flour was also included.

Bread with wheat, rye or barley kernels gave a significantly lower glucose response than the white reference bread. The GI's of the pumpernickel-type products from rye and barley kernels ranged from 50 to 60, which are in fact in the lower range of that reported with pasta[4,5,51-53,55] or pumpernickel bread[29]. These low responses are interpreted in terms of a lowered rate of starch digestion when present in a more intact botanical tissue.

The responses to the oat kernel bread or the barley flour bread were not significantly different from the white wheat reference. The integrity of the oat kernels was partly lost during pre-cooking and baking. Consequently, a disruption of the botanical tissue by milling (barley flour) or heat-treatment (oat kernels) increased glycaemic response. It could also be concluded that the viscous fiber in oats and barley were not effective in reducing glycaemia to bread at the levels tested.

The insulin responses were generally in good agreement with the glucose responses[57]. However, despite the non-effect on glycaemia, the oat kernel bread did significantly reduce insulin response (II 78). Consequently, the effects of naturally occurring viscous fiber in bread were limited to a lowering of insulin response in case of oats. To reduce also the glucose response, bread needs to be enriched with oat fiber[58]. The decrease in GI when adding 45% oat bran to a white wheat bread (GI 72) was of the same magnitude as that observed when adding 80% wheat kernels (GI 73). A bread with improved properties (GI 79) was also obtained by exchanging 30% wheat flour for spaghetti cuts[58].

By creating a more organized structure or by enrichment with viscous fiber, it is thus possible to significantly improve the properties of starch in bread.

Amylose Content / Food Form

A high amylose content is generally considered to lower the metabolic response[35-38]. Possible factors include a higher crystallinity of starch due to incomplete hydration of starch granules at a high amylose content[59], amylose retrogradation[22,24], or formation of amylose-lipid complexes[34,35]. However, no single mechanism has been identified, and the non-consistent data with rice[35-40], make interpretation difficult.

Barley 'Porridges'.   Recently, we evaluated the metabolic responses to barley genotypes, differing in amylose content from 7 to 44%, starch basis (Y. Granfeldt, H. Liljeberg, I. Björck and R. Newman, to be published). The barleys were tested in healthy subjects in the form of porridges made from boiled kernels or boiled flours, respectively.

The impact of amylose/amylopectin ratio was marginal. However, the responses to the boiled kernels were consistently lower (GI 29-37) than to the boiled flours (GI 55-65), again emphasizing the importance of maintaining the botanical tissue. The comparatively low responses also with the flour porridges are not consistent with the high glycaemic response to barley flour bread (GI 93)[57]. This could be due to the fact that the viscous properties of the β-glucans are more effectively developed during boiling of flour than during baking.

These results show that boiled barley products have beneficial properties, particularly in the form of kernels.

Corn Bread 'Arepas'.   In case of corn, certain high-amylose genotypes may contain as much as 75% amylose, starch basis. Products with purified starches from such genotypes have indicated lowered postprandial responses of glucose and insulin[37,38]. However, no data are available on realistic products made from high amylose corn.

The potential of preparing a corn bread from high amylose corn (~75% amylose) was recently evaluated in healthy subjects (Y. Granfeldt and I. Björck, to be published). A corresponding corn product made from common corn (~25% amylose) was included for comparison. The corn was boiled, dried, dehulled and milled into a flour. The flour was heated in a saucepan with water, formed into cakes and finally baked in an oven. The corn bread obtained, so called Arepas, is a common product in Latin America.

A considerable amount of starch in the high-amylose corn bread, about 30% (total starch basis), was enzyme resistant, as judged from in vitro analysis and determination of fecal starch remnants in rats treated with antibiotics[60]. The resistant starch fraction consisted mainly of retrograded amylose. The high-amylose product was therefore tested at two levels, one providing 45g available starch, matching the available starch content in the ordinary Arepa meal, and the other 45g total starch, including resistant starch.

No differences were noted between the high-amylose meals, despite the difference in available starch content, and both these meals produced significantly lower responses than the ordinary Arepa meal. Consequently, the low response to the high-amylose product did not result simply from a lower amount of available starch. The lowered glycaemia probably still arises from amylose retrogradation. A possible mechanism could be that the bulk of amylose in the high-amylose product was retrograded to such an extent that the rate of enzymic digestion was reduced, without a concomitant loss in total digestibility.

The amylose/amylopectin ratio could thus be a parameter of interest when selecting genotypes for human consumption. Possibly, not only the amylose content is of importance, but also the type and extent of processing. Hence, whereas no differences were noted in the rate of in vitro starch hydrolysis between boiled barley flours varying in amylose content from 7 to 33% (starch basis), autoclaving did reduce the rate of amylolysis in case of the high amylose genotype[24]. This was probably mediated by a more prominent amylose retrogradation following autoclaving.

Processing / Botanical Structure (Legumes)

Among starchy foods, legumes generally cause the least accentuated postprandial responses[5]. However, data with beans and lentils indicate an important range in GI, from 25 to almost 90[56]. One interesting topic therefore concerns the possible influence of processing on the properties of legume starch.

In a study of healthy subjects, the postprandial glucose responses to boiled or autoclaved intact red-kidney beans were compared with those to flour based bean products[61]. The flour based products were either processed in a way to maintain intact cells encapsulating the starch[27], or was completely devoid of botanical structure (free starch).

The blood glucose response to the bean products increased in the following order; boiled intact beans, autoclaved beans, flour product with cell-enclosed starch and finally the bean product with free starch. Consequently, boiled intact beans are to be preferred, and excessive heat-treatment or mechanical disintegration will deteriorate properties. Similarly, prolonged drying of boiled lentils has been shown to increase the metabolic responses to nearly the same levels produced with bread[62]. However, even in the complete absence of botanical structure, red-kidney bean starch displayed a comparatively low GI of about 70. This was probably due to the presence of antinutrients e.g. phytic acid or tannins, reducing the rate of enzymic digestion[30,31].

A lentil flour product with remaining cells showed a glucose response similar to that with the bean product devoid of botanical structure[61]. This suggests that the cell wall is a less efficient barrier to amylases in case of lentils. Such differences in e.g. the cell-wall architecture and/or content of antinutrients may influence the potential deterioration of starch properties during processing.

In Vitro Enzymic Ranking of Glycaemic Response

Many enzymic in vitro procedures have been developed with the purpose of predicting glucose and insulin responses to starchy foods[18-20,22,33,47,63]. Amylase incubation is either performed unrestricted or restricted, that is employing dialysis. However, in all of these procedures, products are submitted to mechanical disintegration by e.g. grinding, pasting, homogenization or cutting prior to incubation with amylases. The extent of mechanical disintegration is known to affect the glycaemic response to starchy foods. Hence, milling of rice[28] was accompanied by an important increase in postprandial glucose response. Moreover, type and extent of mechanical treatment also influence the rate of in vitro starch digestion. Not only the hydrolysis level but also the in vitro ranking may be affected by the extent of disintegration[63]. The sample pre-treatment must therefore be considered a critical step, and undue disruption of the food structure should be avoided.

Recently, we described an in vitro procedure based on chewed rather than artificially disintegrated products[64]. The test products are chewed under standardized conditions, expectorated into a beaker and subsequently incubated with pepsin. Finally, the sample is transferred into a dialysis tubing and incubated with added pancreatic α-amylase. Diffusion resistance by viscous fiber may thus be reflected due to the dialysis procedure[65]. Starch degradation products in the dialysate are determined at time intervals, and an hydrolysis index (HI) is calculated with white wheat bread as reference.

A high correlation was obtained with GI when applying this method to a variety of cereal and legume products[64]. Several aspects on food properties were covered, including structure, and the high correlation obtained support the opinion of the rate of starch digestion as a major determinant of glycaemic response. In contrast no differences in rate of in vitro hydrolysis were observed between white bread or pasta[55], nor between white wheat bread and wheat kernel bread[58] following intense mechanical treatment. The in vitro procedure based on chewing can be recommended to facilitate ranking, the advantage being that foods are evaluated with a realistic structure.

A relation between GI and rate of starch digestion does not contradict a relation also to the rate of gastric emptying. Consequently, low GI starchy foods were shown to empty more slowly from the stomach[65]. Digestible solids empty from the stomach when reduced to a particle size of less than 2mm[66]. Due to the presence of α-amylase in the saliva, this is likely to occur more rapidly in case of foods containing readily digestible starch.

4 CONCLUSIONS

It is concluded that the glucose and insulin responses to starchy foods are greatly influenced by factors intrinsic to the raw material as well as by the type and extent of processing. A low glycaemic response is favoured by a low degree of gelatinization, maintaining the botanical structure, creating a dense texture (pasta), the presence of intrinsic components e.g. antinutrients (beans), a high amylose/amylopectin ratio or enrichment with viscous fiber. The importance of a moderate increase in amylose content and naturally occurring levels of viscous fiber is less clear. Finally, the rate of starch digestion was identified as a key determinant of metabolic response. Provided the sample pre-treatment mimicks chewing, in vitro enzymic procedures seem useful to facilitate ranking.

5 REFERENCES

1. A. Dahlqvist and B. Borgström, Biochem.J., 1961, 81, 411.
2. M.L. Wahlqvist, E.G. Wilmhurst, C.R. Murton and E.N. Richardson, Am. J. Clin. Nutr., 1978, 31, 1998.
3. P.A. Crapo, G. Reaven and J. Olefsky, Diabetes, 1977, 26, 1178.
4. D.J.A. Jenkins, T.M.S. Wolever, R.H. Taylor, H. Barker, H. Fielden, J.M. Baldwin, A.C. Bowling, H.C. Newman, A. Jenkins and D.V. Goff, Am. J. Clin. Nutr., 1981, 34, 362.
5. D.J.A. Jenkins, T.M.S. Wolever, A.L. Jenkins, M.J. Thorne, R. Lee, J. Kalmusky, R. Reichert and G.S. Wong, Diabetologia, 1983, 24, 257.
6. D.J.A. Jenkins, T.M.S Wolever and A.L. Jenkins, Diabetes Care, 1988, 11, 149.
7. A.W. Thorburn, J.C. Brand and A.S. Truswell, Med. J. Austr., 1986, 144, 580.

8. F.R.J. Bornet, D. Costagliola, S.W. Rizkalla, A. Blayo, A.-M. Fontvieille, M.-J. Haardt, M. Letanoux, G. Tchobroutsky and G. Slama, Am. J. Clin. Nutr., 1987, 45, 588.

9. T. Pollare and H. Lithell, Forsk.Praktik, 1989, 21, 65.

10. P. Ducimetiere, E. Erschwege, L. Papoz, J.L. Richard, J.R. Claude and G. Rosselin, Diabetologia, 1980, 19, 205.

11. D.J.A. Jenkins, T.M.S. Wolever, G. Buckley, K.Y. Lam, S. Giudici, J. Kalmusky, A.L. Jenkins, R.L. Patten, J. Bird, G.S. Wong and R.G. Josse, Am. J. Clin. Nutr., 1988, 48, 248.

12. J.C. Brand, S. Colagiuri, S. Crossman, A. Allen and A.S. Truswell, Diabetes Care, 1991, 14, 95.

13. D.J.A. Jenkins, T.M.S. Wolever, J. Kalmusky, S. Giudici, C. Giordano, G.S. Wong, J.N. Bird, R. Patten, M. Hall, G. Buckley and J.A. Little, Am. J. Clin. Nutr., 1985, 42, 604.

14. D.J.A. Jenkins, T.M.S. Wolever, J. Kalmusky, S. Guidici, C. Giordano, R. Patten, G.S. Wong, J.N. Bird, M. Hall, G. Buckley, A. Csima and J.A. Little, Am. J. Clin. Nutr., 1987, 46, 66.

15. P. Leathwood and P. Pollet, Appetite, 1988, 10, 1.

16. J.C. Brand, S. Holt, C. Soveny and J. Hansky, Proc. Nutr. Soc. Aust., 1990, 15, 209.

17. A.W. Thorburn, J.C. Brand and A.S. Truswell, Am. J. Clin. Nutr., 1987, 45, 98.

18. D.J.A Jenkins, H. Ghafari, T.M.S Wolever, R.H. Taylor, A.L. Jenkins, H.M. Barker, H. Fielden and A.C. Bowling, Diabetologia, 1982, 22, 450.

19. D.J.A. Jenkins, T.M.S. Wolever, R.H. Taylor, H. Ghafari, A.L. Jenkins, H. Barker, M.J.A. Jenkins, Br. Med. J., 1980, 281, 14.

20. P. Snow and K. O'Dea, Am. J. Clin. Nutr., 1981, 34, 2721.

21. P. Collings, C. Williams and I. Macdonald, Br. Med. J., 1981, 282, 1032.

22. F.R.J. Bornet, A-M. Fontvieille, S. Rizkalla, P. Colonna, A. Blayo, C. Mercier and G. Slama, Am. J. Clin. Nutr., 1989, 50, 315.

23. K. Kayisu and L.F. Hood, J. Food Sci., 1979, 44, 1728.

24. I. Björck, A-C. Eliasson, A. Drews, M. Gudmundsson and R. Karlsson, Cereal Chem., 1990, 67, 327.

25. I. Björck and M. Siljeström, J. Agric. Food Sci., 1992, 58, 541.

26. P. Würsch, S. Del Vedovo and B. Koellreutter, Am. J. Clin. Nutr., 1986, 43, 25.

27. J. Tovar, A. de Fransisco, I. Björck and N.-G. Asp, Food Struct., 1991, 10, 19.

28. K. O'Dea, P.J. Nestel and L. Antonoff, Am. J. Clin. Nutr., 1980, 33, 760.

29. D.J.A. Jenkins, T.M.S. Jenkins, A.L. Jenkins, C. Giordano, S. Guidici, L.U. Thompson, J. Kalmusky, R.G. Josse and G.S. Wong, Am. J. Clin. Nutr., 1986, 43, 516.

30. J.H. Yoon, L.U. Thompson and D.J.A. Jenkins, Am. J. Clin. Nutr., 1983, 38, 835.

31. L.U. Thompson, J.H. Yoon, D.J.A. Jenkins, T.M.S. Wolever and A.L. Jenkins, Am. J. Clin. Nutr., 1983, 39, 745.

32. M. Tamir and E. Alumot, J. Sci. Food Agric., 1969, 20, 199.

33. P. Colonna, J.-L. Barry, D. Cloarec, F. Bornet, S. Gouilloud and J.-P. Galmiche, J. Cereal Sci., 1990, 11, 59.

34. J. Holm, I. Björck, S. Ostrowska, A.-C. Eliasson, N.-G. Asp, K. Larsson and I. Lundquist, Starch/Stärke, 1983, 35, 294.

35. M. Goddard, G. Young and R. Marcus, Am. J. Clin. Nutr., 1984, 39, 388.

36. B. Juliano and M. Goddard, Qual. Plant Hum. Nutr., 1986, 36, 35.

37. K.M. Behall, D.J. Scholfield and J. Canary, Am. J. Clin. Nutr., 1988, 47, 428.

38. J.M.M. Amelsvoort and J.A. Weststrate, Am. J. Clin. Nutr., 1992, 55, 712.

39. P. Srinivasa Rao, J. Nutr., 1971, 101, 879.
40. L.N. Panlasigui, L.U. Thompson, B.O. Juliano, C.M. Perez, S.H. Yiu and G.R. Greenberg, Am. J. Clin. Nutr., 1991, 54, 871.
41. N.A. Blackburn, J.S. Redfern, H. Jarjis, A.M. Holgate, I. Hanning, J.H.B. Scarpello, I.T. Johnson and N.W. Read, Clin. Sci., 1984, 66, 329.
42. B. Flourie, N. Vidon, C.H. Flourent and J.J. Bernier, Gut, 1984, 25, 936.
43. D.J.A. Jenkins, T.M.S. Wolever, A.R. Leeds, M.A. Gassull, P. Haisman, J. Dilawari, D.V. Goff, G.L. Metz and K.G.M.M. Alberti, Br. Med. J., 1978, 1, 1392.
44. D.J.A. Jenkins, T.M.S. Wolever, R. Nineham, R. Taylor, G.L. Metz, S. Bacon and T.D.R. Hockaday, Br. Med. J., 1978, 2, 1744.
45. E. Gatti, G. Catenazzo, E. Camisasca, A. Torri, E. Denegri and C.R. Sirtori, Ann. Nutr. Metab., 1984, 28, 1.
46. T.M.S. Wolever, Am. J. Clin. Nutr., 1990, 51, 72.
47. J. Holm, I. Björck, N.-G. Asp, L.-B. Sjöberg and I. Lundquist, J. Cereal Sci., 1985, 3, 193.
48. J. Holm, I. Lundquist, I. Björck, A.-C. Eliasson and N.-G. Asp, Am. J. Clin. Nutr., 1988, 47, 1010.
49. S.W. Ross, J.C. Brand, A.W. Thorburn and A.S. Truswell, Am. J. Clin. Nutr., 1987, 46, 631.
50. B.W. Arends, Med. Klin., 1987, 82, 277.
51. T.M.S. Wolever, D.J.A. Jenkins, J. Kalmusky, C. Giordano, S. Giudici, A.L. Jenkins, L.U. Thompson, G.S. Wong and R.G. Josse, Diabetes Care, 1986, 9, 401.
52. D.J.A. Jenkins, T.M.S. Wolever, A.L. Jenkins, R. Lee, G.S. Wong and R. Josse, Diabetes Care, 1983, 6, 155.
53. Y. Granfeldt, I. Björck and B. Hagander, Eur. J. Clin. Nutr., 1991, 45, 489.
54. G.B. Haber, K.W. Heaton, D. Murphy and L.F. Burroughs, Lancet, 1977, 2, 679.
55. Y. Granfeldt and I. Björck, J. Cereal Sci., 1991, 14, 47.
56. P. Würsch, Wld. Rev. Nutr. Diet., 1988, 67, 1.
57. H. Liljeberg, Y. Granfeldt and I. Björck, Eur. J. Clin. Nutr., 1992 , 46, 561.
58. J. Holm and I. Björck, Am. J. Clin. Nutr., 1992, 55, 420.
59. P. Colonna and C. Mercier, Phytochem., 1985, 24, 1667.
60. Y. Granfeldt, A. Drews and I. Björck, Starch bio-availability in Arepas made from ordinary or high-amylose corn: Content and gastrointestinal fate of resistant starch. J. Nutr., 1992, (submitted).
61. J. Tovar, Y. Granfeldt and I. Björck, Effect of processing on blood glucose and insulin responses to legumes. J. Agric. Food Chem., 1992 (in press).
62. D.J.A. Jenkins, M.J. Thorne, K. Camelon, A.L. Jenkins, A.V. Rao, R.H. Taylor, L.U. Thompson, J. Kalmusky, R. Reichert and T. Francis, Am. J. Clin. Nutr., 1982, 36, 1093.
63. T.F. Schweizer, S. Reimann and P. Würsch, Lebensm. Wiss. u Technol., 1989, 22, 352.
64. Y. Granfeldt, I. Björck, A. Drews and J. Tovar, Eur. J. Clin. Nutr., 1992, 46, 649.
65. T.M.S. Wolever and D.J.A. Jenkins, J. Plants Foods, 1984, 4, 127.
66. J. Mourot, P. Thouvenot, C. Couet, J.M. Antoine, A. Krobicka and G. Debry, Am. J. Clin. Nutr., 1988, 48, 1035.
67. J.H. Meyer, H. Ohashi, D. Jehn and J.B. Thomson, Gastroent., 1981, 80, 1489.

RESISTANT STARCH: MEASUREMENT IN FOODS AND PHYSIOLOGICAL ROLE IN MAN

H.N. Englyst, S.M. Kingman and J.H. Cummings

DUNN CLINICAL NUTRITION CENTRE, 100 TENNIS COURT RD,
CAMBRIDGE CB2 1QL, UK

1 INTRODUCTION

Starch is the only plant polysaccharide that is susceptible to breakdown by human digestive enzymes, and as such it constitutes an important food resource for populations all over the world. Pancreatic amylase is present in the small intestinal lumen in large amounts, generally regarded to be in excess of requirements. It was partly because of this that it was assumed for many years that the digestion of starch was always fully complete within the small intestine[1,2,3]. However, more recently it has become evident that the rate of starch digestion *in vivo* is highly dependent on crystallinity and food form[4,5] and that starch from many sources is incompletely digested in the small intestine[6,7,8,9,10]. Starch that escapes digestion enters the colon where, through fermentation by the colonic microflora, it may influence large gut physiology.

2 FACTORS AFFECTING STARCH DIGESTION

Dietary starch is degraded in the small intestine by the action of α-amylase. However, certain inhibitory factors reduce the rate at which starch is hydrolysed and absorbed *in vivo*, thus delaying the appearance of glucose in blood after a meal. For some foods, hydrolysis by amylase is retarded to such an extent that part of the starch resists digestion and absorption in the small intestine and passes into the colon. Incomplete starch digestion may be attributed to one or more of the following principal causes:

a) physical entrapment. Starch may be physically entrapped within a cellular or multi-cellular structure which prevents contact with amylase, e.g. starch within a whole cereal grain[11,12,13], or in pasta[14].
b) resistant starch granules. Raw starch may occur in a granular form in which the crystal structure is particularly resistant to amylase, e.g. starch granules in banana and potato[9,10,15].
c) retrogradation. Cooked starch, once cooled, may contain regions where the starch chains (mainly amylose) have crystallised into a configuration that is highly resistant to pancreatic amylase[16].

The factors given above are related to the starchy food itself and may be described as intrinsic factors. Other intrinsic factors that have been shown to affect α-amylase activity *in vitro* include amylose-lipid complexes[17], native α-amylase inhibitors[18], and non-starch polysaccharides which may have a direct effect on enzyme activity[19]. However, it is not clear to what extent these factors affect the digestibility of starch *in vivo*. In addition to these intrinsic factors, starch digestion is influenced by factors related directly to the consumer, which may be called extrinsic factors. Extrinsic influences include the degree of chewing[20], concentration of amylase and amount of starch in the gut, transit time through the small intestine[21], and the presence of other food materials in the digesta[22].

3   NUTRITIONAL CLASSIFICATION OF STARCH

In order to clarify the differences observed in starch digestibility, Englyst and Cummings[23] introduced a nutritional classification of starch, which has subsequently been modified[24,25] and is shown in Table 1. For nutritional purposes starch may be classified into three main types. Rapidly digestible starch (RDS) is starch that is likely to be rapidly and completely digested and absorbed in the small intestine of man. Slowly digestible starch (SDS) is starch that is likely to be completely digested in the small intestine, but at a slower rate. Resistant starch (RS) is starch that is likely to resist digestion in the small intestine and become available for fermentation in the large intestine. Most starchy foods will contain starch belonging to two, if not all three, categories of starch, but in very different proportions. Cooked starchy foods that have an open structure, for example bread and potatoes, contain mostly RDS. In contrast foods that have a dense or protected particulate structure, for example beans and spaghetti, contain substantial amounts of SDS. The amount of RS present in a food will depend on the origin and form of starchy food and on the conditions, if any, under which it has been processed. Particularly high levels of RS are found in raw bananas and potatoes, and in canned beans.

Table 1   Classification of starch for nutritional purposes

Type of starch	Example of occurrence	Probable digestion in small intestine
Rapidly Digestible Starch (RDS)	Freshly cooked starchy food	Rapid
Slowly Digestible Starch (SDS)	Most raw cereals	Slow but complete
Resistant Starch (RS)		
1. Physically inaccessible starch	Partly milled grains and seeds	Resistant
2. Resistant starch granules	Raw potato and banana	Resistant
3. Retrograded starch	Cooled, cooked, potato, bread and cornflakes	Resistant

The three principal causes of incomplete starch digestion are reflected in the three subfractions of RS. Physically inaccessible starch such as that in whole or cracked grains, seeds and legumes is designated RS_1; resistant starch granules are RS_2, and resistant retrograded starch (mainly retrograded amylose) is RS_3. The amounts of RDS, SDS and RS, plus the three RS subfractions in foods can be measured using an *in vitro* method that has recently been developed to accompany this classification[26].

If the value obtained for RDS is expressed as a percentage of the total starch present, a starch digestion rate index (SDRI) is obtained. This index gives a clear indication of the ease with which the starch in a given food is likely to be assimilated in the body. A high value indicates starch that is likely to give a large post-prandial glucose peak, whereas a low value indicates starch that is likely to give a slow release of glucose into the bloodstream. In a recent study of 62 starchy foods (Veenstra and Englyst, unpublished), the SDRI was shown to correlate positively with the physiologically determined glycaemic index[27]. Another index that can be determined by the *in vitro* technique is rapidly available glucose (RAG). The RAG value differs from the SDRI in that it takes into account the water and free sugars content of the food, as well as the digestibility of the starch component. It is measured as the sum of RDS, free glucose and the glucose part of sucrose and is expressed as g/100g of food as eaten. RAG is therefore a measure of the total amount of glucose that is likely to be liberated rapidly into the bloodstream from a given portion of food. It gives a more realistic measure of a food's glycaemic potential for those who need to monitor blood glucose carefully. Both SDRI and RAG may therefore become useful tools for dietitians and diabetic patients in the planning of therapeutic diets.

4 DEFINITION OF RESISTANT STARCH

Resistant starch is defined physiologically as starch that escapes digestion in the small intestine. However, this definition is unsatisfactory from an analytical perspective because it describes a fraction, the magnitude of which is highly dependent on physiological factors. On the other hand, RS cannot be defined in chemical terms because starch resistance is determined primarily by physical factors. Therefore, at the EURESTA (European Flair Concerted Action on Resistant Starch) meeting (Wageningen, June 1990) it was agreed: "that any starch which escapes digestion in the small intestine should be considered as resistant starch (RS). However, it was also agreed that RS has to be defined by a specific *in vitro* technique, which gives an index of the potential of dietary starch to escape digestion in the small intestine". This approach allows the development of *in vitro* techniques by which starchy foods can be classified and compared according to their potential digestibility in man. As part of the EURESTA project a method has been developed for the measurement of RDS, SDS and RS[26]. This method is described briefly below.

5 MEASUREMENT OF RAPIDLY DIGESTIBLE (RDS), SLOWLY DIGESTIBLE (SDS) AND RESISTANT STARCH (RS) FRACTIONS

Figure 1 shows the procedure by which RDS, SDS and RS are measured. The principle of the method is that of controlled enzymic hydrolysis under conditions chosen to reflect the rate and extent of starch digestion found in man. The starch is hydrolysed to dextrins by pancreatic amylase, and further degraded to glucose by the action of amyloglucosidase. The resulting glucose is measured colorimetrically using glucose oxidase.

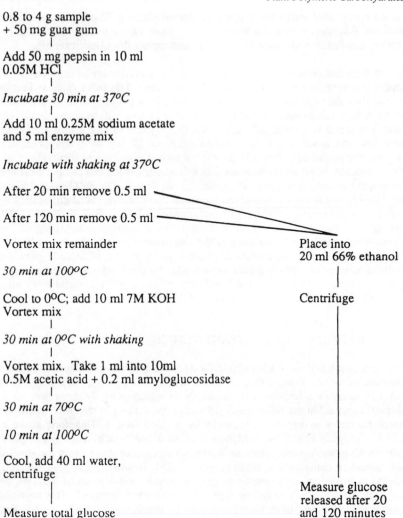

0.8 to 4 g sample
+ 50 mg guar gum
|
Add 50 mg pepsin in 10 ml
0.05M HCl
|
Incubate 30 min at 37ºC
|
Add 10 ml 0.25M sodium acetate
and 5 ml enzyme mix
|
Incubate with shaking at 37ºC
|
After 20 min remove 0.5 ml
|
After 120 min remove 0.5 ml
|
Vortex mix remainder   Place into
| 20 ml 66% ethanol
30 min at 100ºC
| |
Cool to 0ºC; add 10 ml 7M KOH   Centrifuge
Vortex mix
|
30 min at 0ºC with shaking |
|
Vortex mix. Take 1 ml into 10ml
0.5M acetic acid + 0.2 ml amyloglucosidase |
|
30 min at 70ºC
| |
10 min at 100ºC
| |
Cool, add 40 ml water,
centrifuge |
| Measure glucose
| released after 20
Measure total glucose   and 120 minutes

Figure 1   Measurement of rapidly digestible, slowly digestible and resistant starch
              fractions

The starchy food sample is analysed in the form in which it would be eaten. It is
first minced to imitate chewing, incubated at 37˚C with a pepsin/hydrochloric acid solution to mimic the gastric phase of digestion, and then suspended in sodium acetate
buffer at pH 5,2. An enzyme solution containing pancreatin, amyloglucosidase and in-
vertase is added to the suspension, and the sample incubated in a shaking water bath at
37˚C. Glass balls are included in the sample tube to facilitate the physical degradation
of the food.

After 20 minutes incubation, a sample of the suspension is taken into 66% ethanol which prevents further enzyme action. The amount of glucose present in this portion, when corrected for free glucose, represents RDS. After a further 100 minutes incubation, the suspension is sampled in the same way. The glucose measured in this portion, when corrected for free glucose and RDS, represents SDS. Any starch that remains unhydrolysed after 120 minutes incubation with pancreatin and amyloglucosidase is RS.

Once the controlled enzyme incubation is complete, the sample is subjected to more vigorous treatments designed to disperse any unhydrolysed starch. A boiling step, which disperses ungelatinised starch, is followed by agitation with potassium hydroxide which disperses retrograded starch. During these treatments particles continue to be disrupted by the mechanical action of the shaking water bath and the glass balls. A portion of the sample solution thus dispersed is finally neutralised and further incubated with amyloglucosidase. The amount of glucose measured in this portion represents total starch. RS is then calculated as the difference between the total starch and the sum of RDS and SDS. The method gives reproducible values that are determined by factors intrinsic to the food, but includes a mincing step to make the measurements reflect the effect of standard chewing.

6   VALUES FOR RDS, SDS AND RS IN FOODS

Table 2 gives values for RDS, SDS and RS, expressed as a percentage of total starch, in some starchy foods. When expressed in this way, the value for RDS is equivalent to the starch digestion rate index (SDRI). Rapidly available glucose (RAG) values are also reported.

The starch in bread, breakfast cereals and potato measures largely as RDS, indicating that it is gelatinised and readily accessible to pancreatic amylase. Most of these rapidly digested foods also contain a small amount of resistant starch in the form of retrograded amylose formed during cooling. In contrast, ungelatinised starch in raw white wheat flour, raw potato and raw banana flour is much less susceptible to hydrolysis as shown by the high proportions of SDS and RS measured in these products. In the raw state, starch is present in granules in a partly crystalline form which can be of the A, B, or C type as described by Katz[28]. Cereal foods have granules with the A crystal structure, which tend to be more readily hydrolysed than those of the B structure found in banana and potato. Raw cereal flours are therefore high in SDS rather than RS which is the principal fraction in raw potato and banana starches. Biscuits made from wheat, potato and banana starches, in the absence of water, retain some of the granular structure of the raw flour and are therefore also slowly digested or partially resistant to digestion.

Freshly cooked spaghetti, peas and cooked and cooled beans have low SDRI values and contain relatively high levels of SDS and RS. In these foods, although the starch is largely gelatinised, it is less easily hydrolysed because intimate contact between amylase and its substrate is inhibited by the structure of the food. The thick cell walls of beans and lentils have been shown to be a barrier to starch hydrolysis in these foods[29]. In haricot beans cooked to a minimum acceptable softness, starch hydrolysis is inhibited sufficiently that up to 40% of the starch measures as RS (Table 2). This is partly due to physical inaccessibility of the starch, and partly to inadequate gelatinisation and dispersion of the starch. However, if the starch has been completely gelatinised and the cellular structure of the legume seed has been softened by food processing, as is the case for canned baked beans, the main part of the starch is measured as RDS and the RS present is mainly in the form of retrograded amylose.

Table 2   Rapidly digestible, slowly digestible and resistant starch fractions in some common starchy foods, expressed as a percentage of the total starch in the food.

Food	Total starch	% of Total starch			RAG
	(g/100 g dry matter)	RDS (SDRI)	SDS	RS	(g/100 g wet wt)
White flour	81	49	48	3	45
White bread	77	90	9	1	42
Wholemeal bread	60	93	6	1	32
Spaghetti (freshly cooked)	79	52	42	6	13
Digestive biscuits (wheat)	48	77	21	2	47
Cornflakes	78	94	2	4	81
Shredded Wheat	71	93	6	-	68
Porridge oats	65	88	9	3	58
Raw potato	79	2	9	89	<1
Boiled potato (hot)	74	96	1	1	21
Boiled potato (cold)	79	83	11	6	17
Potato/wheat biscuit	55	42	31	27	30
Banana flour	74	4	20	76	6
Banana/wheat biscuit	48	39	22	38	26
Peas (freshly cooked)	20	60	10	25	5
Lentils (cooled)	54	44	41	16	8
Baked beans (canned)	39	81	9	10	44
Haricot beans (boiled 40 min, cooled)	45	18	42	40	4

The RAG value is low for low-sugar dry foods in which the starch is digested very slowly, e.g. biscuits made from banana and wheat flours (banana/wheat biscuits, Table 2), and for foods with a high water content, e.g. potato, even if the starch component is rapidly digestible. Foods that are both wet and slowly digested, e.g. spaghetti and lentils, have particularly low RAG values.

7   COMPARISON OF IN VITRO AND IN VIVO DATA FOR RS

Although an analytical technique for the measurement of RS cannot take account of all extrinsic factors, it is important from the perspective of nutritional research that it yields results that estimate, as closely as possible, those obtained *in vivo*. RS values obtained by the method described above compare favourably with ileal excretion of starch by ileostomists. In a recent study[30] involving 9 healthy volunteers who had undergone surgery to remove their large bowel because of ulcerative colitis, subjects were fed a polysaccharide-free diet together with specially formulated biscuits containing RS from wheat, maize, banana or potato. The biscuits contained from 0.3 to 15 g of RS/100g (either RS_2 or RS_3) as determined by the *in vitro* method. Effluent from the subjects was collected frequently over a 24 hour period, and the amount of starch still present after digestion was determined. For each type of biscuit the mean recovery of starch was close to 100% of RS fed, as shown in Table 3, although there was considerable variation amongst the subjects (74-132% recovery), suggesting the influence of extrinsic factors. In a previous ileostomy study[8] recoveries of starch from white bread, cornflakes and raw oats were 1.5, 3.7 and 1.3 g/100g respectively - values that compare favourably with values of 1.2, 2.9 and 1.6 g/100g for similar products measured by the *in vitro* method described.

Table 3   Recovery of starch in ileal effluent from ileostomists consuming biscuits containing resistant starch

Diet	RS Fed (g)	Starch recovered (g)		Mean (%)
		Mean	Range	Recovery
Wheat biscuit	0.3	0.3	0.1 - 0.4	100
Wheat RS biscuit	8.5	9.0	8.1 - 10.5	106
Maize RS biscuit	8.5	8.6	6.4 - 10.4	101
Potato biscuit	11.7	13.7	12.5 - 15.5	114
Banana biscuit	15.0	13.7	11.1 - 16.6	91

Estimates of starch digestion have also been made by other techniques. Stephen et al.[6] made direct measurements in healthy volunteers using an intubation technique by which samples were aspirated from the ileum after two different meals. The amount of starch resisting digestion from the two meals, which contained rice, banana and potato were 6% and 9% by this method.

8   QUANTITY OF STARCH ENTERING THE COLON, BASED ON IN VITRO MEASUREMENTS OF RS

Values for RS measured using the *in vitro* method described in this paper can be used to calculate the amount of starch, from a given meal, that has the potential to resist digestion in the small intestine. Using data generated in our laboratory from the analysis of individual starchy foods, we have calculated the RS content of three daily menus including the starchy foods shown below:

Menu 1	Menu 2	Menu 3
Shredded Wheat, 40 g	Cornflakes, 40 g	Cornflakes, 40 g
White bread, 60 g	Softgrain bread, 90 g	Banana (slightly green), 150 g
Hot baked potato, 200 g	Canned ravioli, 200 g	Canned potatoes (hot), 150 g
Chilli beans, 30 g	Peas, 100 g	Lentil soup, 200 g
Parboiled rice, 150 g	Bean & chick pea salad, 100 g	Home-cooked beans, 100 g
Cake, 50 g	Potato salad, 100 g	White spaghetti, 100 g
	Digestive biscuit, 55 g	Potato & wheat biscuit, 50 g

All three menus contain 146 g of starch, but have different contents of RS and exhibit different digestion rate characteristics. Menu 1 consists mainly of foods that are well digested in the small intestine. It has a weighted SDRI of 84, and of the starch present only 2.3 g (1.6%) is resistant to digestion. Menu 2 includes some foods that have undergone high temperature food processing or that are known to be poorly digested in the small intestine. This menu has a weighted SDRI of 80 and RS content of 9.4 g (6.4% of total starch). Menu 3 has been devised as an extreme example of RS consumption from (mainly) normal foods, and has a weighted SDRI of only 53. Of the starch present, 36.8 g (25%) is resistant to digestion. Menu 3 demonstrates the potential for delivery of fermentable carbohydrate into the colon from ordinary foods. However, the value given can vary significantly according to food processing conditions, and extrinsic factors can be an important influence on the amount of starch entering the colon for individuals.

It is clear from the data shown in Tables 2 and 4 that, through food processing, the digestibility of starchy foods can be controlled in such a way that large quantities of RS are delivered to the colon. Biscuits containing 30% RS from raw potato starch have been prepared on an experimental scale in our own laboratory. If shown to be beneficial, food technology of this sort could be exploited.

9   PHYSIOLOGICAL ROLE OF RS

Once it enters the colon, RS is available for fermentation by the colonic microflora. The main products of fermentation are the gases carbon dioxide, hydrogen and methane, and the short chain fatty acids (SCFA) acetate, propionate and butyrate which are rapidly absorbed. Butyrate is the preferred fuel for metabolism by the colonic epithelial cells, and much of it is therefore utilised close to its site of production. Butyrate may also benefit the colon in that it acts as a differentiating agent and reduces the propensity to tumour formation[31]. The remaining SCFA are transported to the liver where propionate is cleared and metabolised. Only acetate is released in appreciable amounts into the peripheral circulation, where it is utilised as an energy source by muscle. The energy available to the body from SCFA produced during starch fermentation is up to 60% of the energy that would have been available if the starch had been digested and absorbed in the small intestine[32]. SCFA absorption also stimulates sodium and water absorption, making an important contribution to water and electrolyte homeostasis in the colon. The molar ratios of SCFA produced during fermentation are substrate dependent[33]. In comparison with NSP substrates such as arabinogalactan, xylan and pectin, the fermentation of starch produces higher proportions of butyrate.

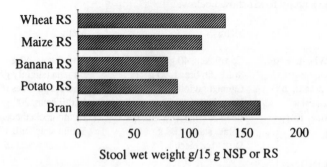

Figure 2   Effect of different resistant starches on stool weight

Data concerning the impact of RS on bowel function are only now emerging. In a study of the effects of different types of RS on faecal weight in healthy volunteers[34] it was shown that on average RS has a laxative effect that is approximately 40% that of wheat bran. However, considerable variation in the bulking effect was seen between the different types of RS (Figure 2). Excretion of faecal dry matter and nitrogen was greater when the subjects were eating RS than when they were on a starch-free or rapidly digestible starch diet, indicating the proliferation of colonic bacteria. No consistent effects on transit time have been observed.

10 CONCLUSIONS

Resistant and slowly digestible starch are emerging as highly important components of the diet. The amounts of RS reaching the colon from many types of food now appear to be greater than those of NSP, and its role as a substrate for colonic fermentation is probably much more important than was hitherto imagined. The individual contributions of NSP and RS in determining faecal weight, transit time and the general health of the colon in normal dietary situations are still uncertain, but with the advent of methods for the detailed analysis of these components in foods, it should soon be possible to make realistic estimates of starch available for fermentation in the large intestine. The rate and extent of starch digestion is highly dependent on crystallinity and physical form, and there is increasing evidence that food processing is a major factor in controlling the digestibility of starchy foods.

11 REFERENCES

1. A. Dahlqvist and B. Borgstrom, Biochem. J., 1961, 81, 411.
2. I.T. Beck, Am. J. Clin. Nutr., 1973, 26, 311.
3. G.M. Gray, 'Physiological effects of food carbohydrates', (Eds. A. Jeanes and J.Hodge), American Chemical Society, Wahington D.C., 1975, pp. 181.
4. K. Hermansen, O. Rasmussen, J. Arnfred, E. Winther and O. Schmitz, Diabetalogia, 1986, 29, 358.
5. K.W. Heaton, S.N. Marcus, P.M. Emmett and C.H. Bilton, Am. J. Clin. Nutr.,1988, 47, 675.
6. A.M. Stephen, A.L. Haddad and S.F. Philips, Gastroenterology, 1983 85, 589.
7. R.W. Chapman, J. Sillery and D.R. Saunders, Gut, 1984, 25, A1158.
8. H.N. Englyst and J.H. Cummings, Am. J. Clin. Nutr., 1985, 42, 778.
9. H.N. Englyst and J.H. Cummings, Am. J. Clin. Nutr., 1986, 44, 42.
10. H.N. Englyst and J.H. Cummings, Am. J. Clin. Nutr., 1987, 45, 423.
11. P.A. Crapo, G. Reavon and J. Olefsky, Diabetes, 1977, 26, 1178.
12. K. O'Dea, P.J. Nestel and L. Antonoff, Am. J. Clin. Nutr., 1980, 33, 760.
13. S. Wong, K. Traianedes and K. O'Dea, Am. J. Clin. Nutr., 1985, 42, 38.
14. P. Colonna, J.-L. Barry, D. Cloarec, F. Bornet, S. Gouilloud and J.-P. Galmiche, J. Cereal Sci., 1990, 11, 59.
15. H. Fuwa, T. Takaya and Y. Sugimoto, 'Mechanisms of Saccharide Polymerisation and Depolymerisation', (Ed. J.J. Marshall), Academic Press, New York, 1980, pp. 73.
16. H.N. Englyst, H.S. Wiggins and J.H. Cummings, Analyst, 2, 107, 307.
17. J. Holm, I. Bjorck, S. Ostrowska, A.-C. Eliasson, N.-G. Asp, K. Larsson and J. Lundquist, Starch/Stärke, 1983, 35, 294.
18. R. Shainkin and Y. Birk, Biochem. Biophys. Acta, 1970, 221, 502.

19. G. Dunaif and B.O. Schneeman, Am. J. Clin. Nutr., 1981, 34, 1034.
20. N.W. Read, I.M. Welch, C.J. Austen, C. Barnish, C.E. Bartlett, A.J. Baxter, G. Brown, M.E. Compton, K.E. Hume, I. Storie and J. Worldling, Br. J. Nutr., 1986, 55, 43.
21. R.W. Chapman, J.K. Sillery, M.M. Graham and D.R. Saunders, Am. J. Clin. Nutr., 1985, 41, 1244.
22. T.M.S. Wolever and D.J.A. Jenkins, 'CRC Handbook of Dietary Fiber in Human Nutrition', (Ed. G.A. Spiller), CRC Press, Boca Raton, 1986, pp. 87.
23. H.N. Englyst and J.H. Cummings, 'Cereals in a European Context', (Ed. I.D. Morton), Ellis Horwood, Chichester, 1987, pp. 221.
24. H.N. Englyst and J.H. Cummings, 'New Developments in Dietary Fiber', (Eds. I. Furda and C.J. Brine), Plenum Press, New York, 1990, pp. 205.
25. H.N. Englyst and S.M. Kingman, 'Dietary Fiber', (Eds. D. Kritchevsky, C. Bonfield and J.W. Anderson), Plenum Press, New York, 1990, pp. 49.
26. H.N. Englyst, S.M. Kingman and J.H. Cummings, Eur. J. Clin. Nutr., 1992 (in press).
27. D.J.A. Jenkins, T.M.S. Wolever, R.H. Taylor, H. Barker, H. Fielden, J.M. Baldwin, A.C. Bowling, H.C. Newman, A.L. Jenkins and D.V. Goff, Am. J. Clin. Nutr., 1981, 34, 362.
28. J.R. Katz, Bakers Weekly, 1934, 81, 34.
29. J. Tovar, A. de Francisco, I.M. Björck and N.-G. Asp, Food Struct., 1991, 10, 19.
30. H.N. Englyst, S. Kingman and J.H. Cummings. Gastroenterology, 1992, 102, A550.
31. Y.S. Kim, D. Tsao, A. Morita and A. Bella, 'Colonic Carcinogenesis: Falk Symposium 31', (Eds. R.A. Malt and R.C.N. Williamson), MTP Press, Lancaster, 1982, pp. 317.
32. K.L. Blaxter, 'The Energy Metabolism of Ruminants', Hutchinson, London, 1969.
33. H.N. Englyst, S. Hay and G.T. Macfarlane, FEMS Microbiol. Ecol., 1987, 95, 163.
34. J.H. Cummings, E.R. Beatty, S.M. Kingman, S.A. Bingham and H.N. Englyst, Gastroenterology, 1992, 102, A548.

SOLUBLE DIETARY FIBRE - A USEFUL CONCEPT?

I.T. Johnson

AFRC INSTITUTE OF FOOD RESEARCH - NORWICH LABORATORY,
NORWICH RESEARCH PARK, COLNEY LANE, NORWICH NR4 7UA, UK

1   INTRODUCTION

The dietary fibre hypothesis is now a little over 21 years old. Simply stated, this theory proposes that diets which are rich in plant cell walls are protective against a range of non-infectious diseases prevalent in industrialised Western societies. Dietary fibre, or "non-starch polysaccharides" (NSP), to use the more modern and increasingly favoured term, comprises the skeletal elements of plant tissues and can be defined as "all those polysaccharides and lignin in the diet that are not digested by the endogenous secretions of the human digestive tract"[1]. In the years following its inception the hypothesis has evolved continously so as to accommodate a host of new research findings. As a result of the advances in both cell wall chemistry and biomedical science which have taken place, it is no longer realistic to talk simply of "dietary fibre", without also specifying its chemical composition and physical properties. Nor is it realistic to envisage a simple theoretical scheme to account for the whole range of protective effects proposed for these polysaccharides. In practice a group of related hypotheses now exists to account for the proposed interactions between non-digestible polysaccharides and human health, and many of these have not yet been rigorously evaluated.

The complexity of the problems involved in relating the diverse physical properties of NSP to human pathophysiology has encouraged various attempts at simplification. It has become increasingly common to distinguish so-called soluble dietary fibre from the other components of cell wall polysaccharides. This concept has become familiar to consumers, and has also been taken up and exploited in the development of new products by food manufacturers.

In general, the soluble components of dietary fibre are thought to be protective against "metabolic diseases" such as hypercholesterolaemia and diabetes mellitus. The insoluble components of fibre act primarily upon the function of the colon, and are thought to play a protective role against such diseases as haemorrhoids, diverticulitis and colorectal cancer. In this paper I will briefly describe the research which has led to these widely accepted beliefs, and I will examine the general validity of the soluble fibre concept.

2 SOLUBLE POLYSACCHARIDES IN THE PLANT CELL WALL

Dietary fibre is a highly complex mixture of polysaccharides derived from a variety of plant tissues. Leafy vegetables, fruits, cereal and legume seeds all contain skeletal elements associated with the vascular, parenchymal and epidermal elements of the plant. Most of these tissues are composed principally of polysaccharides but there are also glycoproteins and lignins in varying quantities. The major types of cell wall polysaccharide are listed in Table 1. Many of these polysaccharides, and in particular the pectic polysaccharides found in fruits, and the beta-glucans which occur in appreciable quantities in certain cereals, are soluble in aqueous media. A general indication of these solubility characteristics is included in the table, but it must be realised that this classification applies to polysaccharides isolated under the appropriate conditions, and it should not be assumed to apply also to the behaviour of these substances during digestion.

Table 1   The principal constituents of dietary fibre and their approximate distribution between insoluble and soluble fractions

Cell wall components

Water insoluble	Water soluble
Cellulose	Pectins
Galactomannan	Arabinogalactans
Xylans	Arabinoxylans
Xyloglucans	Beta-glucans
Lignin	

Analysis of dietary fibre involves the preparation of an alcohol-insoluble residue containing a mixture of polysaccharides, removal of starch by gelatinization, and hydrolysis with amylase, and determination of the remaining polysaccharides. In practice many analytical techniques for dietary fibre only provide a gravimetric estimate of the resistant polysaccharides present in the food, a figure which often includes some resistant starch, much of which is retrograded amylose[2]. The tertiary structure of the skeletal elements within the cell walls is maintained by cross-linking within the matrix. A complete description of the cell wall polysaccharides and phenolics would require a full analysis of the various polysaccharides and their constituent sugars. Even the most sophisticated techniques such as that of Englyst et al.[3] which include analytical procedures for the constituent sugars, do not provide any description of the crosslinks which exist within plant tissues. Our current understanding of dietary fibre in foods is therefore based to a large extent on estimates of total fibre intake, with very little emphasis on the structural organisation that these constituent polysaccharides provide in the intact food. This limitation is particularly significant in relation to the concept of soluble dietary fibre. Many of the pectic components of cell walls are intimately associated with other polysaccharides in the cellwall matrix. Solubilisation can only occur if the crosslinks which determine the structure of these complexes are broken down during digestion. We will return to this problem in a later section, after the biological effects of soluble NSP have been discussed.

Isolated Polysaccharides

The distinction between soluble and insoluble forms of dietary fibre was introduced quite early in the development of the hypothesis. In 1972 Trowell[4] suggested that dietary fibre was protective against ischaemic heart disease and this soon led to studies on the relationship between dietary fibre intake and cholesterol metabolism.

Cholesterol Metabolism. Trowell had proposed that dietary fibre could bind bile acids in the small intestine and thereby prevent their reabsorption, leading to a net loss of cholesterol from the alimentary tract. This would have the effect of reducing the level of LDL-cholesterol in the circulation, a mechanism known to account for the hypocholesterolaemic effect of drugs such as cholestyramine. The most effective bile-salt binding component of dietary fibre is lignin, which suggests that heavily lignified plant tissues such as wheat bran would prove to be the most effective cholesterol reducing sources in the diet. This has consistently been shown not to be the case however. Early studies had demonstrated a hypocholesterolaemic effect of pectin in animals[5], and this was later confirmed for human beings[6]. Animal experiments suggested that the effect depended on the molecular weight of the pectin, and that the mechanism might be mediated by the high viscosity of pectin in the gut contents[7].

The proven hypocholesterolaemic effects of isolated pectin prompted a search for other soluble non-starch polysaccharides with a cholesterol-lowering effect in humans. The storage polysaccharide guar gum, which is obtained from the cluster bean, *Cyamopsis tetragonoloba,* has been shown to be effective, but like pectin, it is relatively unpalatable. Intact foods are of course much more relevant to human nutrition, but palatable sources of soluble polysaccharides are relatively uncommon in western diets. Anderson and colleagues have shown however that diets rich in oats and certain legume seeds consistently reduce plasma cholesterol in hypocholesterolaemic human subjects under controlled metabolic conditions[8,9].

Carbohydrate Metabolism. During the same period another essentially epidemiological theory, namely the argument advanced by Trowell and others that dietary fibre might be protective against diabetes mellitus[10], prompted studies on the effect of isolated polysaccharides on postprandial carbohydrate metabolism. Jenkins and his colleagues added guar gum (16g) and pectin (10g) to solid test meals and observed a flattening of post-prandial glucose tolerance and insulin response in non-insulin dependent diabetes mellitus (NIDDM) patients[11] This work was later extended to liquid test meals containing glucose as used in clinical tests for glucose tolerance. Jenkins and co-workers compared a variety of sources of dietary fibre and demonstrated conclusively that only those which developed a high viscosity in water were capable of reducing the post-prandial glycaemic response. The most effective material studied was guar gum, but the hypoglycaemic response was abolished when the guar was hydrolysed to reduce its viscosity[12]. The quantities of guar gum added to test meals were relatively large but subsequent work has shown that as little as 2.5g of guar incorporated into a liquid test-meal has a substantial hypoglycaemic effect[13].

In parallel with the essentially clinical studies carried out with diabetics and healthy volunteers, physiological studies were also undertaken to establish the mechanism of action of viscous non-starch polysaccharides. One early finding was that under certain circumstances viscous polysaccharides are able to slow the rate at which the stomach empties nutrients into the small intestine, and this led to claims that gastric emptying was the main mechanism whereby guar gum slowed the absorption of glucose[14]. However studies with isolated preparations of animal intestine have demonstrated that viscous polysaccharides increase the resistance to diffusion in the fluid layers surrounding the villi in the small intestine[15]. The delay in gastric emptying following a test meal is probably a less important determinant of nutrient absorption[16], although both factors may have played some role in the delayed absorption of nutrients from meals supplemented with the relatively large quantities of viscous polysaccharides that were used in the early studies.

Intact Foodstuffs

The successful attempts to control postprandial glucose absorption using isolated polysaccharides such as guar gum and pectin led quickly to the development of pharmaceutical products containing a granular form of guar gum. These products were designed to be mixed with food or incorporated into a drink to accompany meals, but they tend to be unpalatable and play only a small role in diabetic therapy. Nevertheless the principle of delaying nutrient absorption continues to interest researchers and clinicians. Jenkins and colleagues have continued to study the factors determining the glycaemic response to foods, and have identified a large number of carbohydrates which are digested and absorbed relatively slowly. Many of these are rich in soluble, if not particularly viscous NSP[17]. In addition to the discovery of more or less natural foods with an intrinsically slow rate of digestion, food manufacturers have succeeded in developing new products rich in soluble NSP[18]. This principle can be extended even to confectionary products such as chocolate, although palatability remains a problem because of the distinctive mouthfeel associated with the high viscosity of such foods[19].

The growing conviction amongst both nutritionists and consumers that plasma cholesterol is an important and a potentially controllable risk-factor for coronary heart disease has stimulated research on the hypocholesterolaemic effects of complex carbohydrate foods. The potential role of oats in this context has proved particularly interesting both because of their commercial importance, and their familiarity to consumers in western industrialised societies. They are also well suited to incorporation into a range of conventional food products such as breakfast cereals and breads. Following the earlier work of Anderson, recent studies have generally confirmed the beneficial qualities of oat bran when incorporated into the diet at relatively high levels, but there have also been some widely publicised reports disputing the conclusion that the hypocholesterolaemic effects observed are due to soluble dietary fibre[19]. Physiological studies with oats have confirmed that the β-glucan fraction of oat fibre becomes solubilised in the small intestine during digestion and confers a high viscosity on the gut contents[20]. Although this effect probably contributes to the relatively low glycaemic index of oats, it is not clear whether it also accounts for their hypocholesterolaemic properties.

The concept of soluble dietary fibre as a separate entity which has beneficial effects in relation to health, and which can be obtained from palatable and familiar foods is now firmly established in the public mind. However the application of this knowledge requires reliable dietary guidelines and probably new products which exploit food technology to the full. Although foods in which the major proportion of NSP is accounted for by soluble polysaccharides such as pectin are rather rare in conventional western diets, most cereal products and many fruits and pulses contain at least some soluble fibre. By judicious selection of such foods a significant daily intake of soluble polysaccharides can be achieved, and some studies have claimed to show a hypocholesterolaemic effect of diets rich in soluble fibre derived mainly from refined wheat products[21]. The exploitation of soluble fibre by dieticians and food manufacturers depends crucially upon a meaningful definition of soluble fibre and reliable methods of analysis.

Analysis of Soluble Fibre

Many methods of analysis of dietary fibre have been developed in the last two decades, and several critical reviews are available[22,23]. Analysis of fibre as NSP is carried out in three main steps comprising: i) preparation of an extractive-free residue by alcohol precipitation. ii) Removal of starch from the residue by gelatinisation and enzyme hydrolysis. iii) Analysis of the residue to determine the composition of the remaining polysaccharides. The 'soluble' fibre present in the sample is usually determined by analysis of the aqueous supernatant after removal of starch and precipitation with alcohol. It is probable that this strictly analytical definition of soluble fibre provides only a rough estimate of the polysaccharides which are solubilised in the small intestine during digestion. Nevertheless it is obviously the latter which is essential to any understanding of the role of soluble dietary fibre in relation to human health. Selvendran and co-workers have shown that the fractionation of cell wall polysaccharides by traditional techniques leads to degradation of pectic substances and hence to loss of soluble fibre[24]. To overcome this problem, non-degradative fractionation techniques have been developed in order to characterise the chemical associations between different cell wall biopolymers, and to more accurately predict the behaviour of the cell wall matrix during digestion[25].

Although cell wall fractionation studies are probably essential for a rigorous understanding of the behaviour of the cell wall matrix during digestion, the demand for a simple and standardised analytical approach to provide at least an approximate estimate of soluble dietary fibre will continue. Considerable caution is needed in applying such techniques, as is well illustrated by the wide variations which have been obtained by different laboratories attempting to characterise soluble fibre for use in physiological studies. Neilson and Marlett[26] analyzed a series of cereals and vegetables using a modification of the method of Theander and Aman[27], and reported that the proportion of fibre recovered in the soluble fraction was usually less than 25%. However more recent reports by Anderson and Bridges[28] and Cummings and Englyst[29] have described levels of soluble fibre of between 30% and 40% for a variety of vegetables and cereal products.

Marlett and co-workers have explored the relationship between analytical methodology and the proportion of total dietary fibre in the soluble and insoluble fractions in some detail[30]. By incorporating a protein digestion step following starch hydrolysis in their standard analytical procedure[26], the yield of soluble fibre was increased. For most foods, a further increase was observed when an additional pepsin step was added (Table 2). Such methodological issues are of considerable importance when an attempt is being made to attribute particular physiological effects to the soluble fibre constituents of foods. For example, Marlett et al.[30] point out that whereas in their studies the proportion of soluble dietary fibre in macaroni varied from 15% to 33%, depending on their method of analysis, all the estimates were considerably less than the figure of 54% reported in the work of Anderson and Bridges[28]. It is quite obvious that, as with any aspect of dietary fibre research, studies on the nutritional or clinical effects of soluble fibre must include a full description of the techniques used to define the material, and existing research must be interpreted with caution.

Table 2   The effect of analytical methodology on measurement of soluble and insoluble dietary fibre in selected foods*

Selected foods	Soluble fibre content (g/100g total fibre)		
	Standard procedure	Protease step	Acid-pepsin
Peas	10	12	16
Rice	6	16	21
Oat bran	40	45	41

Data obtained from Marlett et al. 1989[30].

3   CONCLUSIONS

There is little doubt that certain soluble non-starch polysaccharides can exert significant physiological effects in human beings. Naturally enough, the early studies with isolated polysaccharides have encouraged the search for intact foods with similar biological properties. Foods such as oat bran and legumes contain a large proportion of their fibre content in a fraction which becomes soluble under appropriate analytical conditions, and there is now convincing evidence that such foods have important nutritional characteristics in relation to carbohydrate and lipid metabolism. Nevertheless it must not be forgotten that the attribution of these characteristics to soluble dietary fibre is still largely an assumption, and that even if correct, the mechanism of action remains unclear.

Until the behaviour of soluble cell wall polysaccharides during digestion is more completely characterised it will not be possible to identify precisely which components of this heterogenous group of substances are of importance for human health. The distinction between insoluble and soluble fibre fractions is probably justified for practical purposes, but the terms are of limited significance. The ambiguity of the existing analytical techniques must be taken into account when interpreting research or defining new food products.

4   REFERENCES

1. H. Trowell, D.A.T. Southgate, T.M.S. Wolever, A.R. Leeds, M.A. Gassull and D.J.A. Jenkins, Lancet, 1976, 1, 967.
2. S.G. Ring, J.M. Gee, M. Whitman, P. Orford and I.T. Johnson, Fd. Chem., 1988, 28, 97.
3. H.N. Englyst and J.H. Cummings, Analyst, 1984, 109, 937.
4. H. Trowell, J. Nutr., 1972, 25, 926.
5. T.M. Lin, K.S. Kim, E. Karvinen and A.C. Ivy, Am. J. Physiol., 1957, 188, 66.
6. R.M. Kay, P.A. Judd and A.S. Truswell, Am J. Clin. Nutr., 1978, 31, 562.
7. P.A. Judd and A.S. Truswell, Brit. J. Nutr. 1985, 53, 409.
8. J.W. Anderson, L. Story, B. Sieling, W.-J. Lin Chen, M.S. Petro and J. Story, Am. J. Clin. Nutr.,1984, 40, 1146.
9. J. F. Swain, I.L. Rouse, C.B. Curley and F.M. Snacks, New Engl. J. Med., 1990, 322, 147.
10. H. Trowell, 'Refined Carbohydrate Foods and Disease: Some Implications of Dietary Fibre', Academic Press, London, 1975, p. 227.

11. D.J.A. Jenkins, D.U. Goff, A.R. Leeds et al., <u>Lancet</u>, 1976, <u>ii</u>, 172.
12. D.J.A. Jenkins, T.M.S. Wolever, A.R. Leeds et al., <u>Brit. Med. J.</u>, 1978, <u>1</u>, 1392.
13. H.A. Jarjis, N.A. Blackburn, J.S. Redfern and N.W. Read, <u>Br. J. Nutr.</u>, 1984, <u>51</u>, 371.
14. S. Holt, R.C. Heading, D.C. Carter, L.F. Prescott and P. Tothill, <u>Lancet</u>, 1979, <u>i</u>, 636.
15. I.T. Johnson and J.M. Gee, <u>Gut</u>, 1981, <u>22</u>, 398.
16. N.A. Blackburn, J.S. Redfern, H.Jarjis, A.M. Holgate, I. Hanning, J.H.B. Scarpello, I.T. Johnson and N.W. Read, <u>Clinical Science</u>, 1984, <u>66</u>, 329.
17. D.J.A. Jenkins, T.M.S. Wolever and R.H. Taylor, <u>Am. J. Clin. Nutr.</u>, 1981, <u>34</u>, 362.
18. P.R. Ellis, E.C. Apling, A.R. Leeds and N.R. Bolster, <u>Br. J. Nutr.</u>, 1981, <u>46</u>, 267.
19. I.T. Johnson, J.M. Gee, R.H. Greenwood, G. Wortley, D. Cook and A. Zumbe, <u>J. Sci. Fd. Agric.</u>, 1992, <u>60</u>, 121.
20. E.K. Lund, J.M. Gee, J.C. Brown, P.J. Wood and I.T. Johnson, <u>Br. J. Nutr.</u>, 1989, <u>62</u>, 91.
21. G. Ranhotra, J. Gelroth and H. Bright, <u>J. Food Sci</u>, 1987, <u>52</u>, 1420.
22. D.A.T. Southgate and H. Englyst, 'Dietary Fibre, Fibre-Depleted Foods and Disease' (Eds. H. Trowell, D. Burkitt and K. Heaton), Academic Press, London, 1985, p. 31.
23. R.R. Selvendran, A.V.F.V. Verne and R.M. Faulks, 'Modern Methods of Plant Analysis: New Series' (Eds. H. F. Linskens and J.F. Jackson), Springer-Verlag, Berlin, 1989, <u>10</u>, p. 234.
24. S.G. Ring and R.R. Selvendran, <u>Phytochemistry</u>, 1981, <u>20</u>, 2511.
25. R.R. Selvendran and J.A. Robertson, 'Dietary Fibre: Chemical and Biological Aspects' (Eds. D.A.T. Southgate, K. Waldron, I.T. Johnson and G.R. Fenwick), Royal Society of Chemistry, Cambridge, 1990, p. 27.
26. M.J. Neilson and J.A. Marlett, <u>J. Agric. Food Chem.</u>, 1983, <u>31</u>, 1342.
27. O. Theander and P. Aman, 'The Analysis of Dietary Fibre in Foods', (Eds. E.P.T. James and O. Theander), Academic Press, New York, 1979, p. 215.
28. J.W. Anderson and S.R. Bridges, <u>Am. J. Clin. Nutr.</u>, 1988, <u>47</u>, 440.
29. J. H. Cummings and H.N. Englyst, 'Cereals in a European Context: First European Conference on Food Science and Technology', (Ed. I.D. Morton) VCH Publishers Inc., New York, 1987, p. 188.
30. J.A. Marlett, J.G. Chesters, M.J. Longacre and J.J. Bogdanske, <u>Am. J. Clin. Nutr.</u>, 1989, <u>50</u>, 479.

Industrial Uses

NEW INDUSTRIAL USES OF STARCH

H. Koch, H. Röper and R. Höpcke

CERESTAR RESEARCH & DEVELOPMENT, HAVENSTRAAT 84,
1800 VILVOORDE, BELGIUM

1 INTRODUCTION

The annual production of starch in Western Europe was about 5.6 mio tons in 1990.
About 46% of the starch is used in undegraded polymeric form, either native or modified, in the Food-, Paper-, Corrugating-, Chemical-, Building- and Pharmaceutical industries[1-3].

The remaining 3 mio tons are used as starch hydrolysates, basically in the same industries but with a larger share in the Food sectors: confectionery, soft drinks, fruit preparations and bakery (Figure 1).

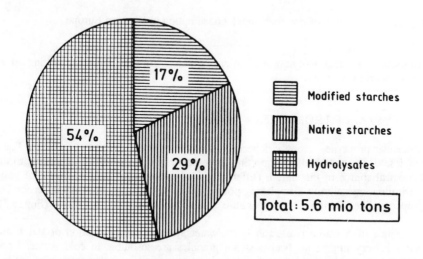

17%

54%

29%

Modified starches

Native starches

Hydrolysates

Total: 5.6 mio tons

Belgium, Denmark, Germany, Finland, France, Great Britain, The Netherlands,
Ireland, Iceland, Italy, Norway, Austria, Sweden, Switzerland, Spain

Figure 1 Consumption of starches in Western Europe 1990

Comparison with the 1981 figures shows a 47% increase of the total starch market, corresponding to a compound growth rate of 4.4%[4]. The higher compound growth rate of the products for the Non-Food areas can be explained by the increased use of starch products in the Chemical and Pharmaceutical sectors which includes fermentation. This sector covers more than 13% of the total starch use and comes closer to the largest starch consumer, the paper industry (Figure 2).

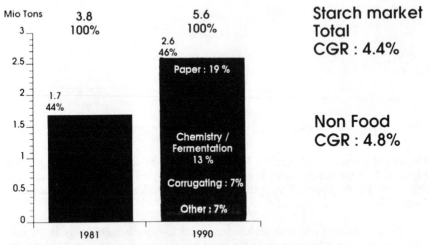

Belgium, Denmark, Germany, Finland, France, Great Britain, The Netherlands, Ireland, Iceland, Italy, Norway, Austria, Sweden, Switzerland, Spain

Figure 2   Evolution of non food starch consumption in Western Europe

This article will deal with new applications of starch, focusing on areas making use of its polymeric structure.

2   STARCH PROPERTIES

Molecular properties strongly depend on the type of starch polysaccharide: the linear amylose or the branched amylopectin. These occur at different ratios according to the biological source of the starch. Differences in the molecular weight, degree of polymerisation, crystallinity, complexing power and retrogradation tendency influence key properties like gelatinisation temperature, solubility and viscosity stability[5,6] (Figure 3).

Since most starch is used in starch/water systems, the interaction of starch and water is very important. Native starch granules are insoluble in cold water. Upon heating, the starch granules swell, which is accompanied by disappearance of the polarisation cross. Gelatinisation, which occurs at a characteristic temperature interval for each starch type, leads to a starch paste in which enlarged granule "ghosts" can still be observed under the microscope (Figure 4). In the chemical-physical sense, there is no complete dissolution of amylose and amylopectin. In particular the amylose easily forms hydrogen bonds and reorganises into linear bundles. This is called retrograda-

		Amylose	Amylopectin
Linkage type	:	α (1→4)	α (1→ 4) and α (1→6)
Molecular weight	:	100.000 - 1. 000.000	1.000.000 - 10.000.000
D.P.	:	≤ 7.000	> 7.000
Morphology	:	Crystalline A, B, V - structure	Amorphous - crystalline
Complexing Power	:	High (blue iodine test)	Low (red iodine test)
Retrogradation	:	High	Low

Figure 3   Properties of starch molecules

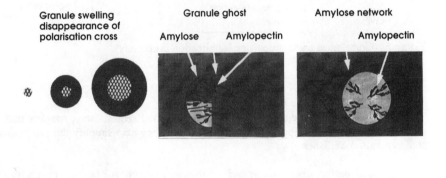

Native starch ⟶ Gelatinisation ⟶ Starch paste ⟶ Gelation ⟶ Starch Gel

Figure 4   Starch gelatinisation / gelation

tion. The pastes become turbid and form stiff gels; this tendency can be used in thickening and texturising applications.

The properties of starch are exploited in a very wide range of industrial applications according to the specific customer demands. There are clear interrelationships between the basic properties like the previously mentioned molecular weight distribution and amylose/amylopectin ratio, and the key properties which can be adjusted by modifications like degradation, esterification or etherification of the starch molecules (Figure 5). In the following application examples we will refer to the key properties of rheology, water binding, adhesive power, gel strength and degradability.

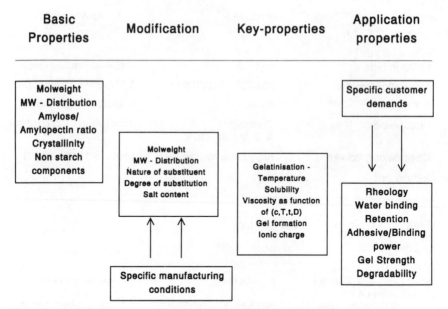

Figure 5   Correlation basic properties / application properties

3   FOOD USES

The first examples for new uses of starch are in the food sector. Here, starches traditionally improve texture, appearance and mouthfeel; they also simplify the processing and preparation of dishes.

Improved performance in comparison to native starch is obtained by chemical modification like crossbonding and substitution. Crossbonding of the starch molecules gives important properties like viscosity stability, inhibited gel formation and resistance to acids as well as to heat- and shear forces. This is achieved by reaction with bifunctional reagents like sodium trimetaphosphate or with adipic anhydride. Esterification or etherification leads to the substitution of a few hydroxyl groups in the starch molecule, resulting in important properties like increased viscosity, enhanced clarity, reduced syneresis and improved freeze/thaw stability. When the substitution is made by hydrophobic reagents, typical emulsifier properties can be achieved. Additional physical modifications like extrusion or agglomeration make the products more suitable for specific applications (Figure 6).

Today more and more ready-to-eat desserts are consumed, many of them are milk-based with a wide range of consistency, from firm-textured through thick spoonable to thin and pourable. Desserts of this type are frequently prepared by Ultra High Temperature (UHT) treatment in order to destroy microorganisms, inactivate enzymes, and extend shelf-life[7,8]. During processing the sensorial properties and nutritional value have to be maintained taking into account the overall economy of the process. It is clear, that due to the high temperatures and shear forces applied during UHT processes, native starch has limited performance. At temperatures around 140°C

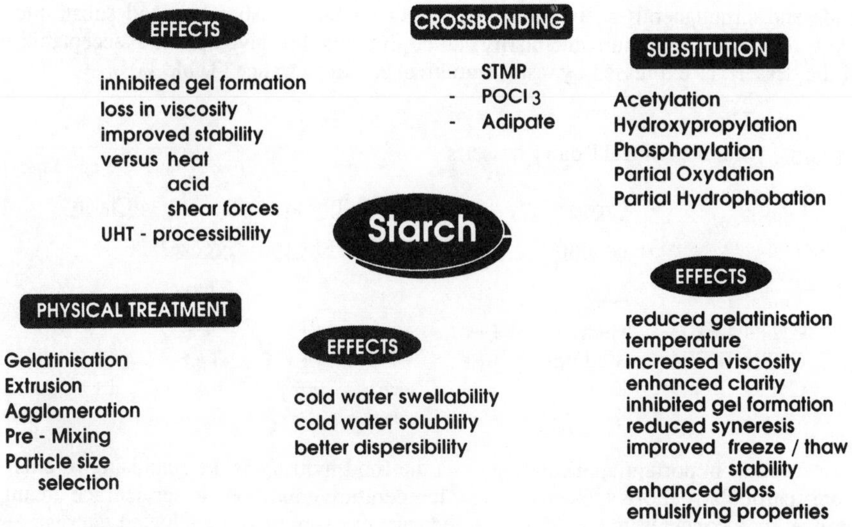

EFFECTS

inhibited gel formation
loss in viscosity
improved stability
versus heat
acid
shear forces
UHT - processibility

CROSSBONDING

- STMP
- POCl₃
- Adipate

SUBSTITUTION

Acetylation
Hydroxypropylation
Phosphorylation
Partial Oxydation
Partial Hydrophobation

PHYSICAL TREATMENT

Gelatinisation
Extrusion
Agglomeration
Pre - Mixing
Particle size
selection

EFFECTS

cold water swellability
cold water solubility
better dispersibility

EFFECTS

reduced gelatinisation
temperature
increased viscosity
enhanced clarity
inhibited gel formation
reduced syneresis
improved freeze / thaw
stability
enhanced gloss
emulsifying properties

Figure 6   Food starch modification

optimum swelling of the starch granules is required during a few seconds of dwell time. Severe shear stress is also a feature of most UHT processes (Figure 7).

	Direct	**Indirect**
Temperatures	ca. 142°C	138-140°C
Time	ca. 5 sec.	5 to 10 sec.

Objectives

1. Destroy micro-organisms
2. Inactivate enzymes
3. Extend shelf-life: uncooled
4. Maintain sensorial properties
5. Preserve nutritional value
6. Maintain low costs for energy, packaging, storage & shipping

Figure 7   Ultra High Temperature (UHT) process

In order to achieve smooth, short texture gels with excellent cold storage stability, careful selection of the raw starch, e.g. waxy maize starch, tapioca starch or blends, is necessary as well as double modification like crosslinking with adipic anhydride and simultaneous acetylation. Depending on the crossbonding and substitution levels the viscosity, texture and clarity can be fine-tuned to give optimum acceptability of the dessert, as expressed by taste, mouthfeel and appearance (Table 1).

Table 1   Starches in UHT dairy desserts

Grade	Cross-bonding level	Stability to:				Clarity
		Heat	Acid	Shear	Freeze/ Thaw	
C*Gel 06203	High	+++	+++	+++	++	++
C*Tex 06304	Medium	+++	+++	+++	+++	+++
C*Flo 06309	Low	++	++	++	+++	+++

Another important application area in the food industry is the stabilising of emulsions, particularly those subjected to low temperature conditions in service: ice cream, orange juice concentrate etc. The food industry has sought and evaluated alternatives to gum arabic for many years. Specific starch derivatives can be obtained by chemical and physical modifications which exhibit the classical properties of emulsifiers: amphipatic structure, i.e. a combination of hydrophobic and hydrophilic character, solubility in both phases with monolayer formation at the phase interface and micelle formation.

Reaction of starch with n-octenyl succinic anhydride gives starch derivates[9] with the required property profile and is the only treatment allowed for emulsifying food starch under current EEC legislation (Figure 8). Waxy corn starch is preferred as base material as it gives clear dispersions with stable viscosity. The reaction takes place under mild alkaline conditions at 30-40°C. The resulting starch bears the necessary hydrophobic substituent but needs additional viscosity reduction which is easily achieved by subsequent extrusion, having the additional advantage that the product becomes cold water soluble.

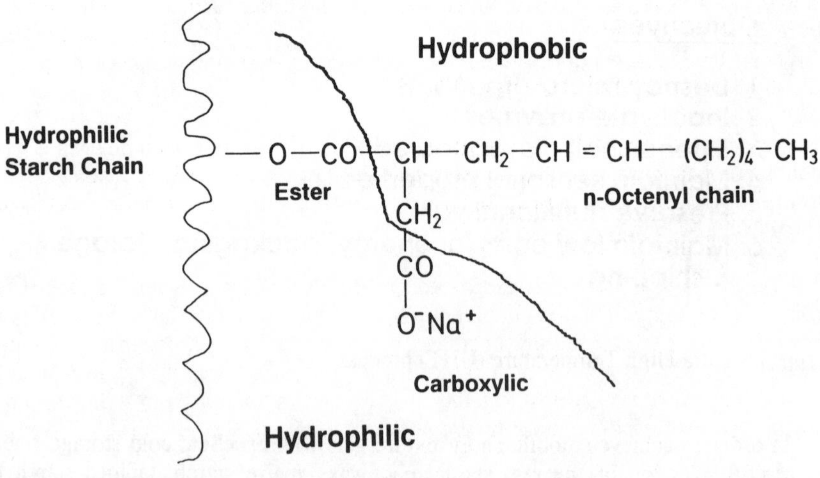

Figure 8   Starch n-octenyl succinate structure

This concept has a wide range of application possibilities such as beverage emulsions and oleoresin emulsions. The encapsulation of flavours gives optimum viscosity stability, using a high percentage of flavouring oils. Cholesterol-free salad dressings are another interesting area as well as cheese analogues. Here, as in dairy creamers, emulsifying starches can be used as an alternative to sodium caseinate[10] (Table 2).

Table 2   Emulsifying starches and their main applications

Application	Function	Alternative to:
Beverage emulsions	Emulsion stabiliser Clouding agent	Gum Arabic
Oleoresin emulsions	Emulsion stabiliser Clouding agent	Gum Arabic
Encapsulated flavours	Encapsulated agent Emulsion stabiliser	Gum Arabic
Salad dressings (cholesterol-free)	Viscosity-stabiliser Eliminate cholesterol	Egg yolk
Cheese analogues	Emulsifier-stabiliser-texturiser	Calcium caseinates
Creamers	Emulsifier	Sodium caseinates

Some years ago, emphasis was on the nutritional value of carbohydrates like starch. Today, there is increasing interest in the functional properties, especially with regard to human health. Food products can play a very important role in the regulation of digestive functions and the prevention of "civilisation" diseases: obesity, diabetes, colon cancer, and cardio vascular disease. The stimulation of natural defence mechanisms is a very important additional role for all food products. Starch-based materials can contribute in this area. Examples are fiber-like resistant starches and fat replacers[11] (Figure 9).

With regard to the latter, the American Heart Association recommends a reduction in fat consumption so that no more than 30% of the total caloric intake is derived from fat. The preferred way to reduce fat consumption is to increase the intake of complex carbohydrates.

Products such as maltodextrins and modified waxy maize starches are already characterised by being able to impart consistency and texture to foods. Low DE potato maltodextrins, i.e. enzymically degraded potato starch with more than 95% higher saccharides, give reversible, meltable gels with water and, therefore, are particularly suitable for partial replacement of fats and oils, while retaining the required appearance, consistency and organoleptic properties of food products[12].

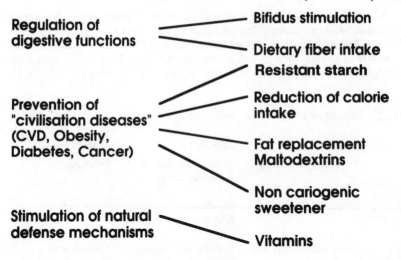

Functional food

The energy value of maltodextrins is about 50% of that of fats and oils. However, considering the usage level, which can be five times lower than for oil and fat, the total caloric reduction is more pronounced and can be as high as 85% calculated on fat. As this calculation does not take into account the energy value of other ingredients, the calorie reduction achieved in food products will be less pronounced and will vary between 15 and 40% (Table 3).

Table 3   Caloric values

Caloric value	Oil/Fat	Potato maltodextrin as is	Potato maltodextrin as 20% gel
kcal/g	9	3.8	0.8
(kJ/g)	(38)	(16)	(3.4)
as % of fat	100%	≤ 50%	15 - 40%

The key property here is the formation of thermoreversible gels at 20% concentration or higher. These exhibit fat-like organoleptic properties. The most interesting application areas include salad dressings where both the traditional hot and a new cold process are available for the preparation of the "Kuli". The calorie reduction for a 50% oil containing standard is about 40% to max. 70%. For a 25% oil-containing salad cream about 55% calorie reduction is feasible. Curry sauces, containing only 15 to 25% oil can be prepared using the maltodextrin together with a pregelatinised modified waxy starch. Reduced fat margarine with a 70% fat reduction can be made by using the potato maltodextrins. Ice creams can be produced with a reduction from 10% fat to 3% or even zero percent. Last but not least, reduced fat Frankfurter sausages and liver paté can be manufactured reducing the standard fat content of 25% to about 11% (Table 4).

Table 4   Fat reduced food formulations

Application	Standard fat / oil content	Reduced fat content	% Caloric reduction
Salad Dressing	50%	25%	40%
	50%	10%	70%
Salad Cream	50%	10%	70%
Curry Sauce	50%	25%	40%
	50%	10%	70%
Margarine	80%	25%	64%
Ice Cream	10%	3%	20%
		0%	30%
Frankfurter /Liver paté	25%	11%	30%

The important role of dietary fibers in human nutrition has been stressed frequently. Similar health benefits are obtained with "resistant starch"[13,14], defined as the sum of starch and products of starch degradation which are not absorbed in the small intestine of healthy individuals; fecal bulk is increased, the formation of short chain fatty acids is induced in the colon, and bile acid is bound, all effects decreasing the risk of colon cancer.

Two possible approaches to "resistant starch" are known which both give the above properties. These are (a) insoluble products consisting of retrograded amylose with a double helical structure and (b) soluble dextrin-like products[15,16] with atypical internal linkages, obtained by rearrangement reactions e.g. through transglycosylation (Figure 10). Both forms are indigestible by the intestinal enzymes and thus only susceptible to fermentation in the colon.

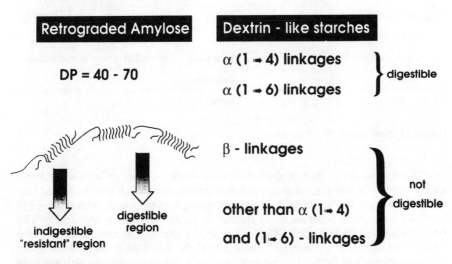

Figure 10 Resistant starch

The potential applications of these "invisible fibers" are in the low caloric food and beverages areas. In comparison to other starch-based products an extra health benefit can be achieved but due to the molecular structure it is difficult to obtain normal functional properties like texture, viscosity enhancement etc. which must be supplied by a modified starch as described earlier. The main research and development activity concentrates on the development of suitable production processes including the enrichment and purification.

4 NON-FOOD USES

Three major areas of non-food use will be illustrated, having in common the positive environmental aspects of the natural raw material starch: biodegradability, biocompatibility and a zero effect on the atmospheric carbon dioxide level.

New binders have been established on the basis of native and modified starches which enable the transformation of powders and dusts into pieces like pellets, bars, briquettes, cylinders and sheets. The different processes can be summarised as follows (Figure 11):

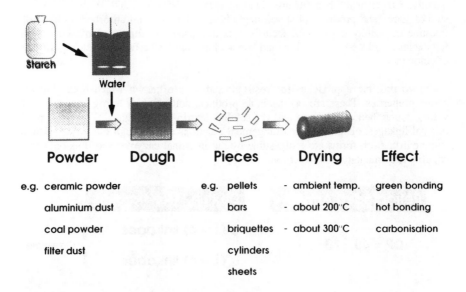

Powder	Dough	Pieces	Drying	Effect
e.g. ceramic powder		e.g. pellets	- ambient temp.	green bonding
aluminium dust		bars	- about 200°C	hot bonding
coal powder		briquettes	- about 300°C	carbonisation
filter dust		cylinders		
		sheets		

Figure 11 New starch based binders

The dry powder or dust of aluminium, coal[17,18] etc. is made up with water and a starch component to a mass with good dough consistency, which then is formed into pieces. The binder has to impart the necessary strength to the pieces, with rapid green bond development in order to avoid fracture in further processing and drying. The specific starch product is selected according to the special requirements of the final product. At drying temperatures of about 200°C the starch component develops its maximum strength in the hot binding step. Special modified starches are also available to form a carbon skeleton by elimination of water at 250-300°C. This effect is used to impart binding power even under red-hot conditions, i.e. carbonisation of the starch.

An example of green bonding is the preparation of ceramics[5] e.g. for electric hot plates, gas or oil-burners of heating systems. By using a cationic starch, a suspension of ceramic mass is flocculated and transformed into a shaped ceramic piece after sucking off the surplus water. The shaped piece can then be finished in the oven. The starch has two effects in this application: improvement of the flocculation/de-watering and ensuring green bond strength.

An example of hot bonding is the pressing of aluminium dust into bars thus facilitating recycling. A dough is prepared from aluminium dust, water and pregelatinised starch. After pressing of bars at normal temperature, they are dried at 200°C which enables transportation and recycling.

Synthetic graphite, e.g. for electrodes, is manufactured by pressing of special coal dust together with special starches and water. The formed pieces are then burnt at 250-300°C under a reducing atmosphere; the starch binder is totally converted into a carbon skeleton inducing perfect binding of the original dust particles. The resulting pieces are covered with a closed wall system and electrically treated for several days at about 3000°C thus forming synthetic graphite (Table 5).

Table 5   Binders and their applications

Green bonding	Hot bonding	Carbonisation
ambient temperature	180 - 200°C	250 - 300°C
Ceramic mass flocculation	Aluminium dust pressing	Synthetic graphite formation
Fly ash compacting	Charcoal briquetting	Coal briquetting
Sinter Waste granulation	Mineral fiber board	Coal dust granulation
Ore dust compacting	Fertilizer granulation	Fire resistant ceramics
	Function of starch	
Starch gives dough structure	Starch gives binding strength	Starch provides carbon skeleton for binding

One of the most important driving forces to use granular, pregelatinised or special carbonising starches is the reduction of air pollution associated with the use of cheaper but problematic binders like bitumen, pith, or sulfite waste waters from paper production. Use of these products causes air pollution which also has a corrosive effect. Increasing restrictions on air contamination imply more and more cost-intensive cleaning of polluted air. This can be avoided by the application of special tailor-made starch products. Their medium term European market potential in the binder sector is estimated to be about 80.000 tons per year (Table 6).

Table 6   Binders and their market volumes

Green bonding	Hot bonding	Carbonisation
Pregelatinized starches	Granular and pregelatinized starches	Special carbonising starches

Medium term market potential in Western Europe

20,000 tpa	50,000 tpa	10,000 tpa

Replacement of problematic binders

e.g. lignin sulfonate waste liquor, bitumen, pitch

There are many application areas where synthetic polymers have replaced traditional starch products. Today, however, due to environmental concerns more and more combinations of starches with synthetics are being investigated and used. A very efficient way to incorporate starches into synthetic polymers is their direct use during the polymerisation reaction thereby exploiting their classical functionality. This type of copolymerisation can either be a type of block-copolymerisation or a type of grafting reaction, using starch as a backbone. In the following examples the production of solid polymers, soluble polymers, and emulsions will be described (Table 7).

Table 7   Starch-containing copolymers

Soluble polymers	Emulsions	Solid polymers
Cobuilders for detergents	Latices for paper coating and carpet back sizing	Superabsorbent polymers for diapers

Starch - based products are incorporated via

grafting or copolymerisation

Very important ingredients of detergents[19] are the so-called "builder" and "cobuilder". These materials complex the calcium and magnesium ions responsible for the temporary hardness of water. If not removed, these ions would precipitate in the form of insoluble carbonates on the fabric as well as form scaling on the metal parts of the washing machine. For a long time phosphates did this job but had adverse environmental impact by causing eutrophication of stagnant waters. In today's environmentally conscious world, combinations of zeolites with polycarboxylates are increasingly used to replace the polyphosphates. Today in Europe already more than 50.000 tons per year of these cobuilders are used together with about 850.000 tons of zeolites; the consumption is still increasing.

The first generation of such polycarboxylic acids are copolymers made from acrylic acid and maleic acid[20,21] (Figure 12).

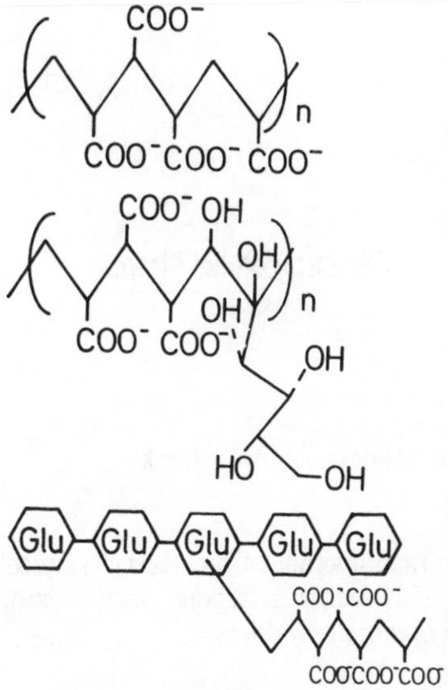

Acrylic acid/Maleic acid Copolymer

DE 29 36 984

Acrylic acid/Maleic acid/Glucose Copolymer

US 4,968,629

Acrylic acid/Maleic acid/Starch Graft Copolymer

EP 0 396 303
EP 0 441 197

<u>Figure 12</u> Polycarboxylate-Copolymers

However, their use is threatened by the fact that they are not biodegradable. They only can be eliminated by adsorption to the activated sludge in waste water treatment plants, which today is still in accordance with, for example, the German law on detergents. The two structures below the acrylic acid / maleic acid copolymer demonstrate two possibilities for the incorporation of carbohydrates into the polymer chain[22] whereby the graft copolymerisation of synthetic side chains onto a starch backbone[23,24] gives products with a better performance. Up to 40% of degraded starches like dextrins can be incorporated without affecting the complexing and dispersing properties for calcium given by the pure synthetic polymer. These products keep, however, the basic concept of using synthetic monomers, and the biodegradability only slightly exceeds the order of magnitude as the amount of incorporated starch.

A new possibility is offered by regarding starch as a polymeric material which can be modified according to requirements. The missing link between the monomeric carboxylic acids like citric acid and the polycarboxylates based solely on starch are the oligomeric oxidised maltooligosaccharides, typically the maltobionic acid. The polymeric polycarboxylate directly based on starch is dicarboxylic starch, a product which retains the polymeric nature of the starch molecule to a large extent (Figure 13).

The carboxylic groups are introduced by oxidative cleavage of the C2-C3 bond in the glucopyranose ring[25,26]. The resulting product is able to form a chelate type structure with bivalent cations. Analytical data confirm the calcium complex stability and direct application trials at the Wäschereiinstitut Krefeld confirm potential use as a cobuilder. Life is not so simple however, several open questions for this application

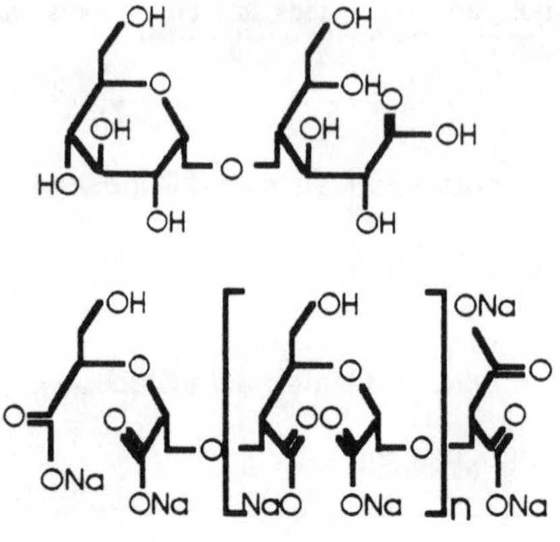

Maltobionic acid

Dicarboxylic starch

Figure 13 Oligo- and polymeric (poly)carboxylic acids derived from starch

remain to be solved. An economic process for the manufacture of dicarboxyl starch is needed, in addition to enhanced biodegradability. This seems to depend on the oxidation process itself and the molecular weight of the obtained products.

Starch/latex mixtures e.g. with acrylates or styrene/butadiene are well known for their use in the paper coating process. Generally they exhibit two problems: a poor wet pick strength due to the water solubility of the starch component and migration of the binder during the drying process, the so-called mottling effect.

When the emulsion polymerisation, exemplified by styrene/butadiene[27], is performed in the presence of a special starch solution, e.g. a dextrin solution, three different possibilities exist:

- polymer particles stabilised by a starch layer which is about the same situation as with a latex starch mixture.
- polymer particles containing starch phases.
- coarse starch particles containing polymer phases.

These three possibilities have been proven by enzymic degradation of the starch-containing dispersions yielding in the first case to spherical polymer particles, in the second case to hollow polymer particles, and in the third to fine polymer particles because the coarse starch particles are degraded[28] (Figure 14).

By using special emulsion polymerisation techniques and varying from the monomer composition either polymer particles containing starch or starch particles containing polymer are obtained. These cases represent very fine distribution of the starch and polymer phases and lead to big improvements in application properties as observed in pilot coater trials:

- total solids contents and viscosities of these dispersions are comparable with commonly used synthetic binders.
- handling is easier than with starch/latex mixtures, because there is no starch cooking necessary.

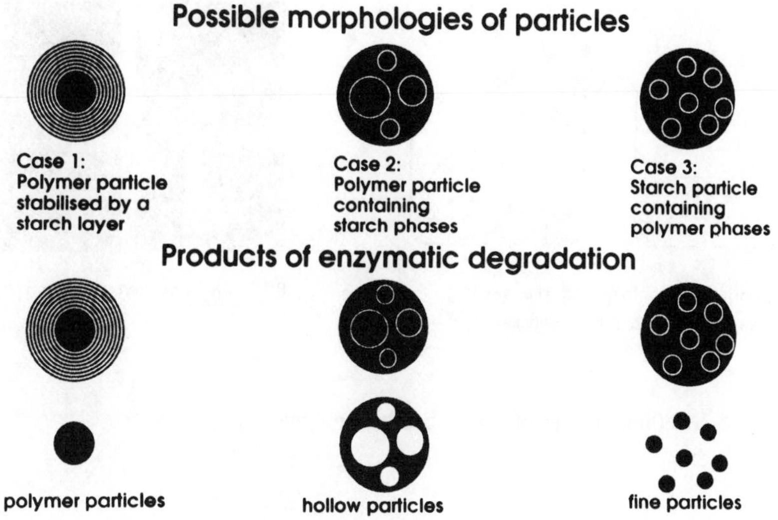

Possible morphologies of particles

Case 1:
Polymer particle
stabilised by a
starch layer

Case 2:
Polymer particle
containing
starch phases

Case 3:
Starch particle
containing
polymer phases

Products of enzymatic degradation

polymer particles

hollow particles

fine particles

Figure 14 Starch-containing emulsions

- high solids coating colours are possible due to controlled rheology.
- the wet pick strength is in the range provided by synthetic binders.
- gloss is improved.
- ink absorption is high.
- mottling and blistering tendency are reduced.

The polymerisation process and the resulting products are covered by a patent application[29] and the products are in the marketing phase.

A very special species of synthetic polymers are the super absorbent polymers, known as SAPs. These are basically polyacrylate salts with a very specific degree of crosslinking[30]. The products can absorb many times their own weight in water, urine or blood and retain it even under pressure. The basic principle is osmosis: aqueous liquids force their way into the "polymeric salt" structure of the SAP until the liquid in and around the polymer reaches osmotic equilibrium. This also explains why the absorption power of SAPs increases with a lowering of the electrolyte content of the aqueous liquid to be absorbed (Figure 15).

By combining the hydrophilic properties of native starch with the absorption properties of the polyacrylate an interesting new species of superabsorbents has been found: partially neutralized acrylic acid is grafted via a special initiator system onto the backbone of the starch molecule[31-35]. By simultaneous crosslinking a kind of interpenetrating network is obtained (Figure 16).

The resulting product consists of small grains which have a raspberry-like primary structure, made up out of many even smaller grains. Thus, the product also keeps its granular structure in the swollen state, a kind of caviar effect.

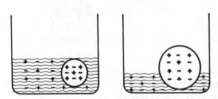

Schematic representation of the swelling
process of Super Absorbent Polymers

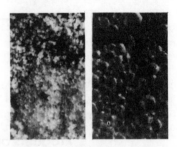

DRYSTAR in swollen and dry state

Figure 15 Swelling process of super absorbent polymers

Schematic molecular structure

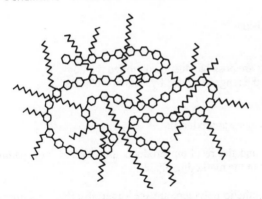

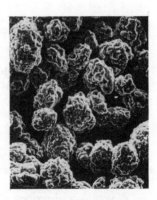

∿∿∿ = polyacrylate chain grafted on starch

○─○ = glucose unit of starch

DRYSTAR - raspberry structure
(100 times magnification)

Figure 16 Molecular structure of super absorbent polymers

These starch-containing products combine high absorption capacity with an extraordinarily fast absorption rate which make them ideal products for the application in baby diapers and adult incontinence products. Additionally, as shown by composting trials, these products show a much higher compostability than it would correspond to the starch percentage they contain. This is regarded as a very important factor for disposable products.

The last part of this article will concentrate on materials based on starch, i.e. starch in different morphological states directly serving different purposes. The common important property of these materials is their biodegradability, which means complete decomposition through the action of living organisms, such as bacteria and fungi, or by enzymes secreted by them into carbon dioxide and water. Examples are described in the area of biodegradable plastics[36-39] comprising composites with oxygenated synthetic polymers like Mater-Bi and plastified starches as well as foamed plastics for packaging chips and clam-shell packaging for fast food. Finally, a new fibrous starch material, Chart-Bi, will be described (Table 8).

Table 8   Polymeric starch-based materials

Biodegradable plastics	Foamed plastics	Fibrous material
Mater - Bi	Packaging chips	Chart - Bi
Plastified	Clam schells	

Functionality

Starch morphology is adapted for specific applications

The driving force for the use of biodegradable polymers is the immense amount of more than 100 million tons of Municipal Solid Waste produced each year in Western Europe. In this MSW, plastics account for 5-10% of the weight but 25-30% of the volume, and volume for landfill, is exactly what is running short (Figure 17).

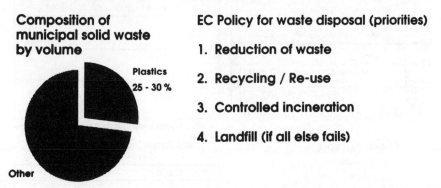

Composition of municipal solid waste by volume

Plastics 25 - 30 %

Other

EC Policy for waste disposal (priorities)

1. **Reduction of waste**

2. **Recycling / Re-use**

3. **Controlled incineration**

4. **Landfill (if all else fails)**

Biodegradation / composting to be accepted as biological recycling

Figure 17 Biodegradable plastics

The first draft for an EC directive for tackling the waste problem gives the priorities in decreasing order as shown in Figure 17.

In this respect, biodegradation is nothing else than biological recycling and, consequently, fully in line with this policy. Agricultural raw materials, tranformed into plastics can be re-transformed, after use, into compost, biomass and carbon dioxide, thus representing a completely closed system.

A material which fits this purpose is Mater-Bi from Novamont[40-47] having the following characteristcs:

- starch and natural additive content higher than 60% on a weight basis.
- synthetic part is a hydrophilic copolymer of vinyl alcohol, able to interact with starch. This component is also biodegradable.
- interpenetrated or semi-interpenetrated microstructure at molecular level.

The high content of hydrophilic groups guarantees significant resistance to light, fats, oils and non-polar solvents as well as excellent antistatic and ink-printing characteristics. The products do not dissolve in cold or hot water but swell according to the type of additive and transformation process used. The mechanical properties at 23°C and 55% relative humidity are within the range of LDPE and HDPE. Industrial products exist for film blowing, injection molding, blow molding, thermoforming and extrusion.

The biodegradation speed of products made with Mater-Bi is comparable to that of paper (Figure 18).

Biodegradability tests

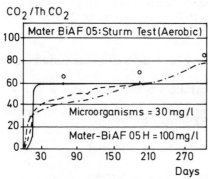

Sc as Test and biodegradation in water : Mater - Bi AF1 OU

Biodegradation time (days)	Weight loss (%)	
	SCAS°	LAKE WATER
75	70	-
210	77	-
460	83	-
660	not found	79

SCAS = Semicontinuous Activated Sludge

Soil Burial Test: Mater - Bi A FO5H (Fertec Method)

Days	Weight loss
7	> 39 %
21	> 50 %
49	> 55 %

Figure 18 Biodegradability tests

The biodegradation behaviour has been analysed by using[48]:

- respirometric aerobic tests, based on CO_2 evolution, i.e. a modified Sturm test: 8% of the theoretical CO_2 has been recovered when stopping the test after 300 days.

- semicontinuous activated sludge testing: complete disappearance after 660 days. Lake water testing gave a 79% weight loss in water.
- soil burial test gives more than 55% weight loss after 49 days.

Similar products are being developed[49,50].

Without the addition of other synthetic polymers starch can also be processed under controlled high pressure, temperature and moisture in an extruder or by injection molding to give biodegradable thermoplastic products[51,52] (Figure 19).

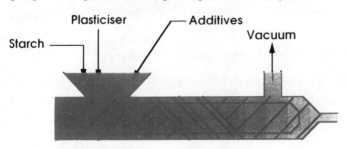

Process conditions : T : 100 - 150° C

p : 20 - 70 bar

starch is : - destructurized

- blended with plasticizer + additives

- partly degraded

(- reacting with additives)

Figure 19 Plastification of starch

The plastification is related to the glass transition temperature and this depends very much on the moisture content. Starch with less than 20% moisture, extruded at 140-170°C with addition of low volatility plasticisers like glycerol or sorbitol gives thermoplastic products which can be used for film preparation of injection molding. A major problem is still the product stability caused by eventual migration of the plasticiser. Several institutes and companies, including us, are working on the optimisation of these products.

Several companies are also working on foamed materials on the basis of starch for the replacement of conventional synthetic packaging materials[53], whereby foaming is either done by blowing agents or by special extrusion conditions (Table 9): Biopac[54] for example uses a modified waffle-baking technology to produce foamed starch articles for trays-packaging. The material has good heat insulation and shock adsorption properties. Potential substitution of polystyrene trays for the fast food sector is under development.

Südstärke[55] is producing loose-fill materials for packaging on the basis of potato starch and Storopack makes similar products on the basis of maize starch. The products have bulk weight as low as 20 kg per m³. A special product on the basis of high amylose starch is manufactured by National Starch. The closed cell structure is advantageous in comparison to the usual open structure. Additionally the product exhibits reasonable ageing stability under normal climatic conditions. All these materials are very water sensitive or readily soluble. The biodegradability is improved due to the increased surface area.

Table 9   Foamed biodegradable plastics

Product and Raw Material	Company	Brandname	Application
Extruded starch foam (Potato starch)	Südstärke (D)	Aeromyl chips	Loose fill
Extruded starch foam (Maize starch)	Storopack (D)	Renatur	Loose fill
Extruded starch foam (hydroxypropylated high amylose starch)	National Starch / American Excelsior (USA)	Eco - foam	Loose fill
Starch foam articles (starch, fibers, additives)	Biopac (A)	Biopac	Trays packaging

As a last example of a new starch based material[56,57], Chart-Bi[58], a fibrid manufactured completely out of maize starch will be mentioned. This product, developed by Novamont, is produced by a special spinning technology. The fibrids are about 1 mm long and contain approx. 60% water (Figure 20). They are white and completely miscible with cellulose fibers. Depending on the type of paper or board, between 3 and 25% can be added during production.

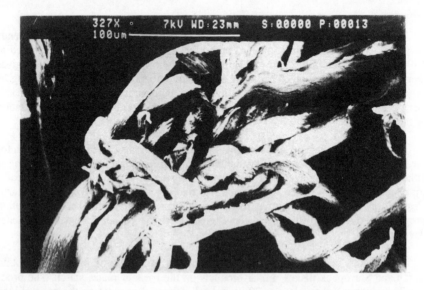

Figure 20 Starch fibrids

The addition of Chart-Bi improves the Scott Bond values and the Dension values, both important strength properties requested by the paper industry. The effect of Chart-Bi is a combination of fiber properties and starch-related binder properties.

First industrial trials are underway at paper manufacturers to provide the necessary feed-back for the final product optimisation. With this product an annually renewable material like starch could replace large parts of cellulose produced in a multiannual cycle.

5   SUMMARY AND OUTLOOK

A variety of new uses for starch has been shown for the Food and Non-Food Sector. Besides new products for convenience food, functional food will attract increasing attention in the coming years offering possibilities for calorie reduction, stimulation of digestion and in general prevention of diseases.

In the Non-Food area alternative binders can contribute to solve environmental problems like air pollution due to the potential replacement of problematic binders like bitumen and pitch.

Starch copolymers and the new starch-based materials are increasingly requested by the consumers due to their environmental friendliness (Figure 21).

The market introduction of these new products is at different phases. Many of the technical questions are not yet completely answered and need further research and development.

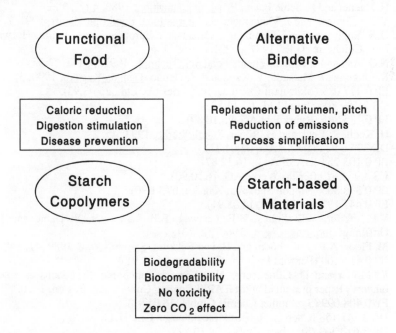

Figure 21   Renewable raw material

178 *Plant Polymeric Carbohydrates*

Depending on their success, which certainly will also be linked to production economics, an additional potential of several hundred thousand tons of starch seems feasible. This would represent a 10-20% increase on the industry's present capacity and provide an additional market for agriculture requiring 100 to 200 thousand hectares of land.

Public opinion is in favour of it!

6 REFERENCES

1. H. Koch and H. Röper, Starch/Stärke, 1988, 40, 121.
2. N.O. Bergh and J.-L. Hemmes, Wochenblatt für Papierfabrikation, 1991, 4, 111.
3. H. Röper, Agro-Industry Hi-Tech, 1991, 2(4), 17.
4. 'Daten zur Stärkenindustrie'. Fachverband der Stärke-Industrie e.V., Bonn, 1992.
5. R.L. Whistler, J.N. BeMiller and E.F. Paschall, 'Starch: Chemistry and Technology', Academic Press Inc., New York, 1984.
6. H. Koch and H. Röper, 'Polysaccharide', Springer, Berlin, 1991, Chapter 8, p. 177.
7. A. Rapaille, J. Vanhemelrij and J. Mottar, Dairy Ind. Int., 1988, 53, 21.
8. A. Rapaille and J. Vanhemelrij, 'Comparative functionality study of modified starches in different UHT processing systems', Paper presented at Food Ingredients Asia, Bangkok, 13.-15.05.92.
9. US 4,035,235 (Anheuser-Busch, Inc., 12.07.77).
10. V. De Coninck and M.G. Fitton, 'n-Octenyl succinate starches. Production and application properties', Paper presented at the 43rd Starch Convention, Detmold, 28-30.04.92.
11. H. Haenel and F. Schierbaum, Ernährung/Nutrition, 1980, 4 (7), 306.
12. V. De Coninck and J. Vanhemelrijck, International Food Ingredients, 1991, 2, 27.
13. H.N. Englyst and J.H. Cummings, 'Cereals in a European context', Ellis Horwood Ltd., Chichester U.K., 1987, p. 221.
14. N.G. Asp and I. Björck, Trends Food Sci. Technol., 1992, 3, 111.
15. K. Ohkuma, I. Matsuda, Y. Katta and Y. Hanno, Denpun Kagaku, 1990, 37, 107.
16. EP 0 477 089 (Matsutani Chemical Industries Co. Ltd., 25.03.92).
17. GB 2 189 806 (Inrad Ltd., 04.11.87).
18. EP 0 135 785 (Ruhrkohle AG, 25.10.89).
19. H. Koch, R. Beck and H. Röper, Starch/Stärke, 1993, 45, 2.
20. DE 29 36 984 (BASF AG, 02.04.81).
21. EP 0 103 254 (BASF A.G., 16.12.87).
22. US 4,963,629 (Grillo-Werke AG, 16.10.90).
23. EP 0 396 303 (Nippon Shokubai Kagaku, 07.11.90).
24. EP 0 441 197 (BASF AG, 14.08.91).
25. V.F. Pfeifer, V.E. Sohns, H.F. Conway, E.B. Lancaster, S. Dabic and E.L. Griffin, jr., Ind. Eng. Chem., 1960, 52, 201.
26. M. Floor, A.P.G. Kieboom and H. van Bekkum, Starch/Stärke, 1989, 41, 348.
27. EP 0 411 100 (Penford Products Company, 06.02.91).
28. K. Möller and D. Glittenberg, 'Novel Starch containing dispersions as coating binders', Paper presented at the TAPPI Coating Conference, Boston, 1990.
29. EP 0 408 099 (Synthomer Chemie GmbH, 16.01.91).
30. DE 3 741 158 (Chemische Fabrik Stockhausen GmbH, 08.03.90).
31. DE 3 613 309 (Starchem GmbH, 22.10.87).
32. EP 0 324 985 (Starchem GmbH, 26.07.89).

33. DE 3 801 633 (Starchem GmbH, 27.07.89).
34. DE 4 014 628 (Starchem GmbH, 14.11.91).
35. EP 0 455 965 (Starchem GmbH, 13.11.91).
36. H. Röper and H. Koch, Starch/Stärke, 1990, 42, 123.
37. Bundesministerium für Forschung und Technologie, 'Untersuchung zum Einsatz bioabbaubarer Kunststoffe im Verpackungsbereich', Forschungsbericht Nr. 01-ZV 8904, Bonn, Aug. 1991.
38. G. Delheye, 'The role of biodegradable plastics in modern waste management', Paper presented at the 43rd Starch Convention, Detmold, 28-30.04.92.
39. Institute for International Research, 'Bioabbaubare Verpackungen - Chancen für Marketing und Entsorgung', Conference papers, Fachkonferenz Mannheim (D), 30.-31.03.92.
40. EP 0 400 531 (Butterfly, S.R.L., 05.12.90).
41. EP 0 400 532 (Butterfly, S.R.L., 05.12.90).
42. EP 0 413 798 (Butterfly, S.R.L., 20.09.90).
43. EP 0 437 561 (Butterfly, S.R.L., 21.02.91).
44. EP 0 436 689 (Butterfly, S.R.L., 24.02.90).
45. EP 0 437 589 (Butterfly, S.R.L., 21.02.91).
46. WO 92/01743 (Butterfly, S.R.L., 06.02.92).
47. WO 92/02363 (Butterfly, S.R.L., 20.02.92).
48. C. Bastioli, V. Bellotti, L. Del Giudice and G. Gilli, 'Mater-Bi: Properties and Biodegradability', Paper presented at the symposium "Environmentally degradable polymers; Technical, Business and Public Perspectives", Lowell, 13-15.08.91.
49. EP 0 404 723 (Warner-Lambert Company, 27.12.90).
50. US 5,095,054 (Warner-Lambert Company, 10.03.92).
51. EP 0 397 819 (I. Tomka, 17.05.90).
52. W. Wiedmann and E. Strobel, Starch/Stärke, 1991, 43, 138.
53. Bayerisches Staatsministerium für Ernährung, Landwirtschaft und Forsten, 'Gesamtkonzept Nachwachsende Rohstoffe in Bayern', Agra-Europe 42/91, 14.10.91.
54. WO 91/12186 (Franz Haas Waffelmaschinen Industriegesellschaft m.b.H., 22.08.91).
55. DE 3 206 751 (Südstärke GmbH, 01.09.83).
56. US 2,902,336 (Avebe 01.09.59).
57. US 4,139,699 (National Starch and Chemical Corporation, 13.02.79).
58. Chart-Bi. Fibrilles d'amidon de mais: Un nouveau materiaux pour la fabrication des papiers et cartons. Novamont Italia S.P.A., Via Principe Eugenio 1/5, I-20155 Milano.

PROPERTIES OF SMALL STARCH GRANULES AND THEIR APPLICATION IN PAPER COATINGS

E. C. Wilhelm

INSTITUTE FOR STARCH AND POTATO TECHNOLOGY, FEDERAL CENTRE FOR CEREAL, POTATO, AND LIPID RESEARCH IN DETMOLD AND MÜNSTER, W-4930 DETMOLD, GERMANY

ABSTRACT

Technologically important physical and chemical properties of small starch granules from rice and amaranth as well as small starch granule fractions from wheat and oats have been studied. These starches had uniform granule sizes in the range 0.5 to 3.0 µm. Shear stress and intrinsic viscosity of the native starch granules suspended in water were investigated to obtain a better understanding of the rheological properties and specific flow behaviour of small starch granules and thus help improve separation techniques using decanter, hydrocyclone, and other centrifugation systems. Due to starch/water interactions and immobilization of water molecules on the large specific surface area of small starch granules, a typical increase in intrinsic viscosity was found compared to those suspensions made with large starch granules. Significant differences in quality characteristics of small granule starches were discovered by microscopic and particle size distribution measurements. Uses of small starch granules include the coating of graphic papers, and the grafting of their starch molecules with synthetic polymers to form starches with improved properties in coatings for paper. Cross-linking reaction techniques on small starch granules using their different reaction behaviour to esterification and other chemical reactions have been investigated especially for small wheat and oat starches granules.

1   INTRODUCTION

Novel raw materials for starch manufacturing processes[1-4] have gained greater importance since high valued by-products can be obtained from these raw materials. Besides this, the starches of these raw materials exhibit special characteristics like distinct granule sizes and granule size distributions. As a raw material, oats contain small starch granules whose flow and viscous properties can be technically exploited. The extraction of oat starch by wet milling is combined with the extraction of fiber, ß-glucans, proteins and lipids which are of particular importance for human nutrition. This made it possible to develop an economically feasible process.

It is already known that small granuled starches, like rice starch, are of special interest for the paper industry[5]. It is therefore understandable that this industry is a major source for the demand of starches featuring special characteristics important for paper production processes as well as for paper quality criteria.

Although various fractionation techniques and extraction processes[6] have been described for separating different oat products by wet-milling, a suitable process for the production of oat starch was not developed until 1987[7]. This process is based on multistage hydrocyclone fractionation and purification techniques.

Fractionating and purification of crude oat starch by means of high performance multistage hydrocyclone techniques and counter current washing with the addition of fresh water to give a primary (A) and a secondary (B) starch fraction are shown in Figure 1. An effective separation of oat starch is possible in 10 mm diameter multi-stage hydrocyclones at elevated g-forces, corresponding to high operating pressures. The hydrocyclone w1 overflow is split into B-starch and residual A-starch, the latter being recycled to the feed of the system. Due to a pressure drop of 8 bar, correspond-ing to a g-force of 8500, fractionation and purification of small starch granules from larger ones is achieved.

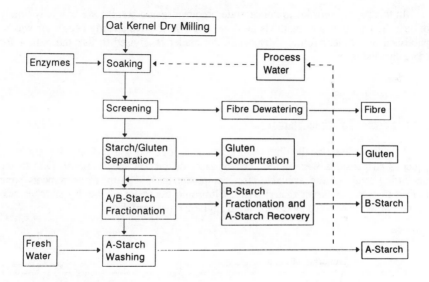

<u>Figure 1</u>   Process for fractionation of oats

The size of oat starch granules is in the range of that of rice starch, but only oat starch can be effectively separated and purified by multistage hydrocyclone techniques. Both starches have narrow ranges of particle size distributions. Whereas rice starch consists of polyhedral granules, oat starch granules are spherical. It is mainly this shape characteristic which results in a better performance of the oat starch in separation and purification processes applying hydrocyclone techniques as compared to rice starch.

For these reasons, starch industries are showing more and more interest in the production of starches which consist of small starch granules even though a lot of ef-fort is still needed to improve the recovery of the starch, its separation from by-prod-ucts, its purification by centrifugation processes as well as its dewatering and drying. The size of starch granules is important in each of these process steps. For example, in many production processes, the main part of the small starch granules fractions is lost due to ineffective separation techniques.

Therefore, as an initial step, we investigated the viscous properties of native, purified small starch granules in aqueous suspensions at about 20-25°C in order to understand the flow behaviour of these particles.

For these investigations the extremely small granule starch from Amaranth[8-9] (A.caudatus) were recovered, by means of an improved dry milling process in combination with a wet extraction process followed by lab scale techniques for the recovery of uniform and very small starch granules. The diameter of the largest granules of the extracted starch was 3.0 μm and its mean diameter, the statistically calculated median value of idealized ball shaped particles, was 1.6 μm. Thus, a starch with a very narrow range of granule sizes was available to which other starches of similar granule sizes, like starch granule fractions from wheat and barley, could be compared in their flow as well as other characteristics, such as intrinsic viscosity and accessability to chemical derivation.

In this paper, in order to concentrate on our research results, we will only briefly outline the materials and methods used as far as these have already been[4,7] or will be published in the near future. Meanwhile the reader is advised to ask the author for more detailed information.

2   RESULTS

Particle Sizes of Various Starches

Starch particle size distributions for amaranth, oat and wheat A-starch granules in aqueous suspensions were determined by laser diffraction spectrometry. Due to the measuring range of the available laser equipment of 1-118 μm, measurements were limited to particles larger than 1.0 μm. Ultrasonic treatment of the dry starches suspended in water was used to disintegrate agglomerated particles.

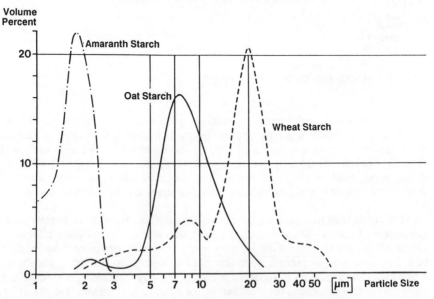

Figure 2   Particle size distribution of amaranth, oat and wheat starches

The measurements have shown that more than 30% of amaranth starch granules were in the range of 0.5-1.2 µm. Figure 2 shows particle size distribution curves of amaranth, oat and wheat starches.

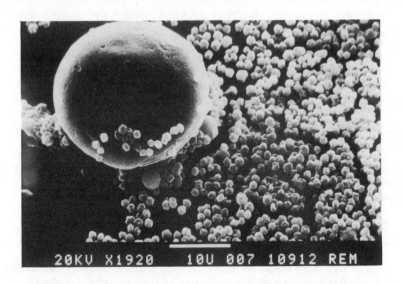

Figure 3   SEM micrograph of amaranth and wheat starches

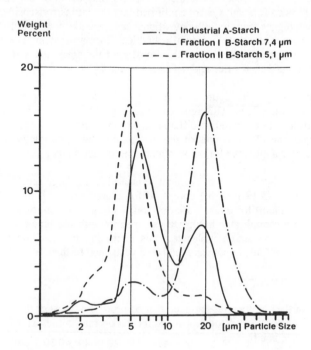

Figure 4   Particle size distribution of wheat A- and B-starch fractions

Fractionation of wheat starch by hydrocyclone and decantation techniques resulted in a significant enrichment of SSG in the overflow. Particle size distribution of two SSG fractions from wheat starch are shown in Figure 4, where they are compared to commercially extracted wheat starch granules of the A-starch fraction. The bimodal distribution showed that only 10-20% of the wheat A-starch was similar in size to oat starch.

To illustrate the shape of amaranth and small wheat starch granules SEM micrographs were taken. Figure 3 shows that the granules of the small granule starch (SGS) of amaranth and small starch granules (SSG) of wheat starch ranged between 0.5-3.0 μm in diameter.

Flow Behaviour of Small Starch Granules Suspended in Water

Aqueous starch suspensions can normally be concentrated in the underflow of hydrocyclones to 43% dry substance (d.s.), but refinement of oat starch in a 12 stage hydrocyclone system could not exceed a maximum concentration of about 30% d.s. This was a significant factor for further investigations.

Differences in Brookfield viscosity of 18% d.s. starch slurries were found for amaranth compared to oat and wheat starches. At low shear rates (10 RPM) a 6.7 times higher viscosity was found for amaranth starch compared to wheat starch, which increased to 16.6 times higher when the amaranth starch suspension was acidified to pH 1. But in alkaline suspensions of pH 11 no change of viscosity was observed.

Detailed and more exact rheological measurements of 43% d.s. starch suspensions with a Physica Rheolab system confirmed that the intrinsic viscosity of SSG and SGS was much higher than that of conventional starches, like corn starch. The investigations were carried out with a coaxial cylindrical Rheolab system. The results demonstrated an extreme increase in intrinsic viscosity for the amaranth starch slurry compared to medium effects for rice and oat starch slurries.

The curves shown in Figure 5 were plotted by down scaling measurements to eliminate thixotropic effects. Differences in viscosity between rice and oat starch slurries were substantial compared to differences for corn and wheat starches. This is in agreement with one of the claims of the oat fractionation technology patent[7] which maintains that it is applicable for oat starch but not for rice starch.

Differences in intrinsic viscosity might be caused by the large specific surface area of amaranth starch (5.19 m^2/cm^3) compared for example to 0.18 m^2/cm^3 for potato starch and 0.49 m^2/cm^3 for maize starch by which a larger amount of water molecules attached to the amaranth starch surface can interfer with the break-up of hydrogen bondings between the starch molecules thus increasing the swelling power of the starch/water-system. This hypothesis shall be investigated further.

Process Technology Aspects

In some of the commercial starch extraction processes, most of the smallest starch granules are lost because of too low a sedimentation rate in the centrifuge by which the separation rate is influenced.

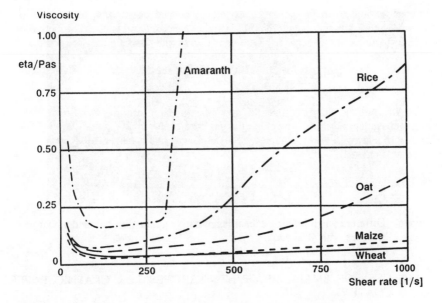

Figure 5   Intrinsic viscosity of aqueous suspensions (43 weight-%) of native starches
as a function of granule size/starch type

The rate of separation depends on:

- hydration of the starch granules which affects their densities and
- agglomeration of the starch granules with proteins and fiber which affect both their
shapes and densities.

Therefore, specific processes for separating small starch granules using decanters
and/or hydrocyclones as separation techniques should be developed. These techniques
should also be combined with the process steps of purification and dewatering of the
starch slurries.

Dewatering of Small Starch Granule Slurries

The concentrating of starch slurries causes an increase in their intrinsic viscosity
due to granule interactions. Thus, the upper concentration limit for slurries of small
starch granules is obtained at lower d.s. concentrations compared to slurries containing
larger starch granules. The dry matter content of dewatered slurries of SSG and SGS
is therefore correspondingly lower than that for their counterpart.

In Figure 6 the different water binding capacities of amaranth, oat and wheat
starches are shown. All contain exactly 52% water, but the wheat starch was a mobile
slurry, and the amaranth starch appeared to be almost dry.

Figure 6   Different water binding capacities of native amaranth-, oat- and wheat
            starches with 48% d.s.

3   USES OF SMALL GRANULE STARCHES FOR GRAPHIC PAPER COATING

Starches for graphic paper coatings should have diameters in the range of 10 μm or
below. They should be reasonably stable to alkali of a pH value of 9.0-9.5 at maximum temperatures of 65°C, and should be fairly resistant to shear stress to which they
are exposed in high speed coating machines.

Because of their very low granule size, which favours the surface coating in the
paper production process, small granule starches could be particularly suitable for the
production of graphic papers since an improved printability and a less glossy surface of
the paper could be achieved. Therefore small granule starches with low water binding
and low swelling capacities were developed by cross-linking the starch molecules of
SSG and SGS. These special features of the derivatized starches are of particular importance, as modern high speed paper coating processes need very short drying periods.

Cross-linking of Small Starch Granules

Cross-linking reagents need some activation with alkali for kinetically controlled
reactions with starches. In a series of reagents tested, sodium trimetaphospate [STMP]
was less effective for bifunctional reactions due to the demand for a high pH-value.
Phosphorus oxychloride [$POCl_3$] and epichlorohydrin [ECH] proved to be more efficient due to the decrease of swelling capacity of small starch granules in alkaline media
at elevated temperatures. Cross-linking power of $POCl_3$ was superior to that of STMP
but difficult to control.

Particle size distribution curves calculated from more than 3 000 single data per
laser diffraction measurement showed statistically significant deviations between the
treated starch granules. An increase in the median diameter D (v, 0.5) was found for
the cross-linked small starch granules when compared to the untreated starch granules.
Several process conditions seem to be responsible for this fact being best recognized in
the case of the SGS extracted from amaranth (Table 1).

<u>Table 1</u>   Particle size of native and cross-linked amaranth-, oat- and wheat starches

Starch	Cross-linked by	Degree of Substituion	D (v, 0.5) [μm]	Spec. Surface Area [m²/cm³]
Oat-SGS	native	-	7.9	0.833
Oat-SGS	Spray dried	-	8.1	0.804
Oat-SGS	STMP	0.01	9.0	0.733
Oat-SGS	POCl₃	-	9.1	0.728
Oat-SGS	ECH	0.07	9.3	0.724
Amaranth-SGS	native	-	1.6	5.194
Amaranth-SGS	STMP	0.01	1.7	4.483
Amaranth-SGS	POCl₃	0.09	2.1	3.552
Wheat-SSG	native	-	6.5	0.976
Wheat-SSG	STMP	0.01	6.8	0.928
Wheat-SSG	ECH	0.08	6.9	0.914

STMP: Sodium trimeta phosphate
POCl₃: Phosphorus oxychloride
ECH: Epichlorohydrin
DS: Degree of substitution
Dv 0,5:   Median value of particle size diameter

<u>Stability of Coatings</u>

The stability of maize, oat (SGS) and wheat (SSG) starches were tested in alkaline coatings at pH 9 using 15% starch plus 85% mineral pigment in comparison to a standard coating with 100% mineral pigment. For a thermal stability test, the Brookfield viscosities of these coatings were measured at increasing temperatures from 25 to 45 and at 65°C, and then cooled down to 25°C. In Figure 7, stable coating viscosities are shown. Maize and oat starch viscosities remained close to that of the standard coating, while the wheat starch coating viscosity increased to 7160 Pa s on final cooling, a value far too high for paper production.

Subsequently, a series of measurements was carried out on starch pigment coatings for papers which were produced by using 30% starch plus 70% mineral pigment. Measurements concerned the most important properties for fine printing papers; abrasion resistance, smoothness and gloss of the paper surface. In Figures 8-10 results are shown comparing calendered to non-calendered paper treated with maize, wheat and oat starch coatings in comparison to a standard coating without starch.

In all papers the abrasion resistance (Figure 8) was stabilized in the calendered form, while smoothness (Figure 9), the very important quality criterion for printing papers, increased to three-fold that of the standard coated paper in the case of oat starch. The gloss of paper surfaces was relatively high for oat starch (Figure 10), however, this is a property no longer desired for many papers. However, glossiness was successfully reduced almost to standard in later trials.

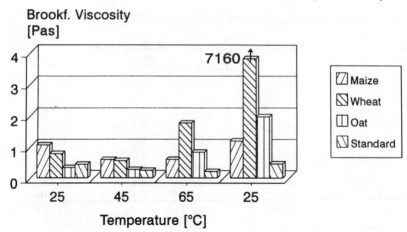

Figure 7   Brookfield viscosity of heated starch coatings; 15% maize, wheat or oat
starches; 85% mineral pigment

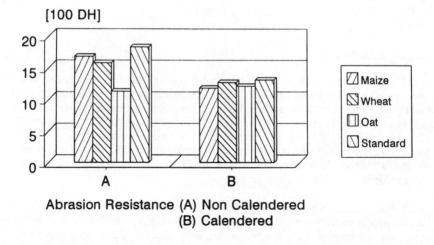

Figure 8   Coated paper measurements, coatings with 30% starch pigment: abrasion
resistance of non calendered (A) and calendered paper (B)

The development of this process for the production of small granule starch coatings is the first application of these starches for the production of smooth but non-
glossy lightweight, coated fine printing papers. It opens up promising perspectives for
the use of SSG and SGS. This is a challenging opportunity for process engineering to
develop new separation and purification techniques in starch extraction from suitable
raw materials.

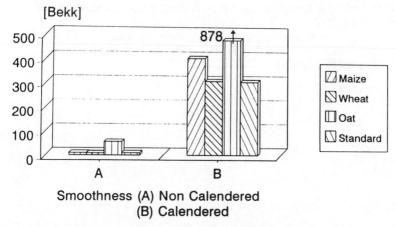

Figure 9   Coated paper measurements, coatings with 30% starch pigment: smoothness of non calendered (A) and calendered paper (B)

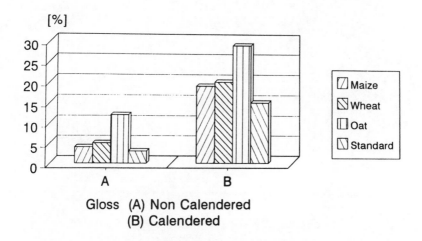

Figure 10 Coated paper measurements, coatings with 30% starch pigment: gloss of non calendered (A) and calendered paper (B)

4   REFERENCES

1. L. Munck, F. Rexen and L.H. Pedersen, Starch/Stärke, 1988, 40, 81.
2. H.U. Woelk, Starch/Stärke, 1981, 30, 397.
3. F. Schierbaum, S. Radosta, M. Richter, B. Kettlitz and C. Gernat, Starch/Stärke, 1991, 43, 331.
4. E. Wilhelm and W. Kempf, Starch/Stärke, 1987, 39, 153.
5. R.L. Whistler, J.N. BeMiller and E.F. Paschall, 'Starch: Chemistry and Technology', Academic Press, Inc., Orlando, Florida, USA, 1984.
6. G.A. Hohner and R.G. Hyldon, U.S.Pat. 4,028,468 (1977).

7. E. Wilhelm, W. Kempf, A. Caransa, P. Karinen and A. Lehmussaari, E.P. 0381872 (1990).
8. L.A. Stone and K. Lorenz, Starch/Stärke, 1984, 36, 232.
9. O. Paredes-López, A. Carabéz-Trejo, S. Pérez-Herrera and J. González-Castaneda, Starch/Stärke, 1988, 40, 290.

TECHNICAL APPLICATIONS OF GALACTOMANNANS

F. Bayerlein

DIAMALT GMBH, GEORG-REISMÜLLER-STR. 32, W-8000 MÜNCHEN 50, GERMANY

ABSTRACT

Galactomannans are mainly found in the endosperm of seeds from the Leguminosae or pea family. Important sources are the seeds of Cyamopsis tetragonoloba for Galactomannan-1,2 (Guar), Cesalpinia spinosa for Galactomannan-1,3 (Tara), Ceratonia siliqua for Galactomannan-1,4 (Locust bean) and Cassia tora/obtusifolia for Galactomannan-1,5 (Cassia gum).

These polysaccharides are used either in their native states or as (carboxyalkyl-, hydroxyalkyl-, trimethylammoniumalkyl-) derivatives to thicken aqueous systems. Their properties depend mainly on their chemical structure, i.e. chain length, availability of cis-OH-groups, steric hindrance, substituents.

Chain length, i.e. molecular weight of the polysaccharide, influences viscosity and the rheological properties of the aqueous solution. Solubility in water depends on the extent of intermolecular hydrogen bonding and whether steric hindrance keeps the chains at such distances from each other that water can penetrate in between and hydrate or dissolve the galactomannan.

The textile industry uses galactomannans, and derivatives thereof, in printing inks and as sizes for textile weaving. The paper industry makes use of galactomannans in paper pulp for retention and thickening in papercoating, and also in printing inks. Galactomannans and their derivatives also play an important role as ameliorating agents in drilling fluids and oil recovery, and in the suspension and flotation of minerals. But, the main field of application is still the use as gelling or thickening agents in the feed and food industries.

1 INTRODUCTION

Increasing amounts of hydrophilic products on the basis of polygalactomannans are used to ameliorate water-based processes or products. The application-range for these additives stretches from food to textiles and from paper to minerals and petroleum. Galactomannans play an important role as ameliorating agents in processes where water has to be thickened, or where hydrophilic materials need to be coated, depressed or suspended.

Furthermore the group of galactomannans presents to the chemist a very surprising and interesting example of how nature achieves variations in water solubility of polysaccharides by the degree of substitution of a relatively insoluble polymannan-backbone.

Some new findings about galactomannans from Cassia tora also known as Cassia obtusifolia are discussed here. These new polysaccharides are easily available and distinguished from others by their specific ability to form gels with polymers bearing carboxyl- or sulfoxyl-groups.

2 ORIGIN, STRUCTURE AND CHEMICAL COMPOSITION OF GALAC-TOMANNANS

Galactomannans are polysaccharides consisting mainly of the monosaccharides mannose and galactose. The mannose-elements form a chain consisting of many hundreds of $(1{\rightarrow}4)$-β-D-mannopyranosyl-residues, with $(1{\rightarrow}6)$ linked α-D-galactopyranosyl-residues at varying distances, dependent on the plant of origin. This structure has been elucidated by extensive work over the past five decades. The main techniques used were methylation, partial hydrolysis, periodate oxidation and specific, enzymic hydrolysis[1].

A backbone of mannose-residues which is substituted to a certain degree by galactose and sometimes also minor amounts of glucose form the galactomannans (Table 1).

Galactomannans are mainly found in the endosperm of seeds from the Leguminosae or pea family although not too many Leguminosae bear endosperm in their seeds. Leguminosae are capable of fixing nitrogen from the air via the root-tubercle bacterium radicicola and therefore do not need artificial fertilizing - in numerous cases an ideal crop for underdeveloped areas or waste land.

Many plants give a "dowry" to the next generation depositing an energy-stock in their seeds, bulbs or nodules in the form of fat or polysaccharides. In some families the polysaccharides occur in the seeds as galactomannans. However, just as wheat starch differs significantly from corn starch, rice starch or starch from any other source there are also remarkable differences between the galactomannans derived from various Leguminosae.

Table 1 gives a simplified scheme (M = Mannose, G = Galactose) of the structure of such molecular chains. The distances between the galactose-substituents are considered to be irregular of course. Some authors suggest that galactose might be linked in accumulated aggregates to the polymannan so that larger sections are left unsubstituted and are more easily accessible to reactive partners[2].

Seed galactomannans appear to have a double physiological function. Firstly, they retain water by solvation and thereby prevent (in regions having arid conditions) complete drying of the seeds which would cause protein denaturation, in particular, the denaturation of those enzymes essential for seed germination. Secondly, the galactomannans serve as food reserves for the germinating seeds.

<u>Table 1</u>   Chemical structure of polymannan and galactomannans

Polymannan

```
- M - M - M - M - M - M - M - M - M - M - M -
```

Galactomannan-1,5 (Cassia-Gum)
from the endosperm of Cassia tora/obtusifolia

```
- M - M - M - M - M - M - M - M - M - M - M -
  |                   |                   |
  G                   G                   G
```

Galactomannan-1,4 (Carubin)
from the endosperm of Ceratonia siliqua

```
- M - M - M - M - M - M - M - M - M - M - M -
  |               |               |
  G               G               G
```

Galactomann-1,3 (Tara-Gum)
from the endosperm of Cesalpinia spinosa

```
- M - M - M - M - M - M - M - M - M - M - M -
  |           |           |           |
  G           G           G           G
```

Galactomannan-1,2 (Guaran)
from the endosperm von Cyamopsis tetragonoloba (Guar plant)

```
- M - M - M - M - M - M - M - M - M - M - M -
  |       |       |       |       |       |
  G       G       G       G       G       G
```

Galactomannan-1,1

```
- M - M - M - M - M - M - M - M - M - M - M -
  |   |   |   |   |   |   |   |   |   |   |
  G   G   G   G   G   G   G   G   G   G   G
```

The use of locust bean galactomannans probably occurred prior to other galacto-mannans. Analyses showed that it had served to thicken the mucilage which was used to glue textile windings around mummies in ancient Egypt. In our time locust bean galactomannan-1,4 (Carubin) came onto the market during the twenties, produced and promoted by companies like Tragasol, Cesalpinia and Diamalt. Guar galactomannan-1,2 (Guaran) joined the club in 1942, although it has been used for centuries in India in food and feed. It was developed during world war II in the USA as the main areas around the Mediterranean sea, in which locust bean trees were growing, had been affected by the war. Thus access to locust beans was lost for the US. General Mills successfully started to promote Guar-gum as a thickener for aqueous solutions.

Subsequently numerous authors searching for such substances, investigated subtropical Leguminosae, amongst others, for galactomannans[2, 3, 4, 5, 6, 7]. This work - and our own investigations - led to a small collection of galactomannans, useful for industrial applications. They differ in their chemical structure mainly by the galactose/mannose ratio and they offer hydrophilic polymers on a natural basis, capable of dissolving or swelling directly in water or with chemical pretreatment.

As mentioned before, galactomannans are polysaccharides consisting mainly of the single sugars galactose and mannose (Table 1). An ideal formula could be drawn for galactomannan-1,5 (Cassia gum) and for galactomannan-1,2 (Guaran), for example (Figures 1,2):

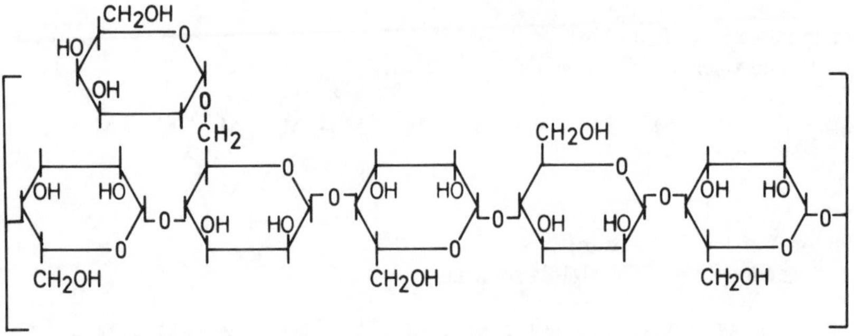

Figure 1   Formula for galactomannan-1,5

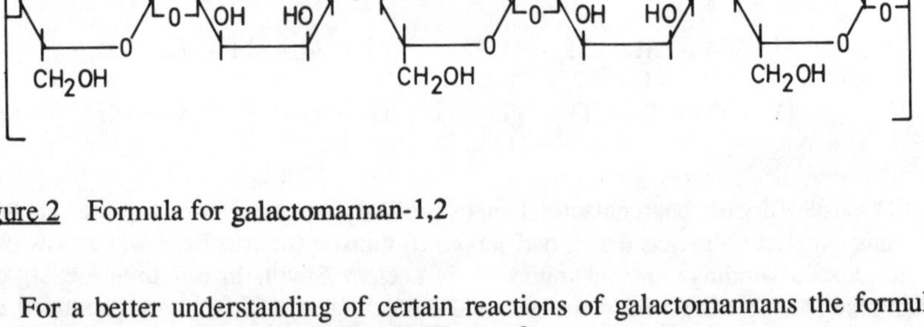

Figure 2   Formula for galactomannan-1,2

For a better understanding of certain reactions of galactomannans the formulas for galactose and mannose are shown in Figures 3.

It is obvious that both sugars are provided with cis-OH-groups. Therefore, an enhanced disposition to form hydrogen bonds between the polymannan-chains is found as long as voluminous neighbouring groups like galactose do not develop too much steric hindrance to prevent the galactomannan-chains from coming too close together and thus prevent the mannose cis-OH-groups from reacting.

GALACTOSE   MANNOSE

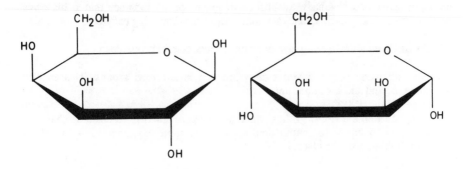

<u>Figure 3</u>   Formulae for galactose and mannose

<u>Physical Properties of Galactomannans</u>

Comparing the different galactomannans of Table 1, it is obvious that, since additional crosslinking via hydrogen bonds goes hand in hand with less solubility, an increase in substitution leads to higher solubility.

Polymannan for example is almost insoluble in cold as well as in hot water and galactomannan-1,5 (Cassia) shows only partial solubility in cold water resulting in a low viscosity. After boiling, the same suspension yields a thick, colloidal solution of high viscosity. This is caused by the thermal breakdown of the inter-molecular hydrogen bonds so that water can penetrate in between the chains leading to hydration of the molecules. So a gelatinisation point is found - not as precise as for starches - but within a distinct temperature range.

Galactomannan-1,4 (locust bean) acts in a similiar way. It does not dissolve in cold water but its solubility in boiling water is already better than that of galactomannan-1,5 due to the slightly higher degree of substitution by galactose side groups compared to galactomannan-1,5.

It is obvious that natural gums do not only consist of homogeneous polysaccharides. They may vary in molecular weight, i.e. chain length, as well as in the galactose : mannose ratio. Therefore the above-mentioned figures are to be seen as average values which can differ for certain crops, portions or fractions.

Locust bean gum seems to be more heterogeneous than guaran[8] and we assume the same for cassia-gum since we also find in this new gum a cold water soluble fraction, a hot water soluble fraction and an insoluble fraction consisting of polymannan. Therefore, if we talk about of guaran, Tara gum, carubin, cassia gum, we are only dealing with average statistical data.

Galactomannan-1,3, Tara gum, with an average galactose/mannose - ratio 1:3 - acts similarly to locust bean gum. However, galactomannan-1,2, guar gum, where on average each second Mannose unit is blocked by Galactose, shows so much steric hindrance and so little hydrogen bonding between the molecular chains that it hydrates instantly. The best solubility is found with galactomannan-1,1 in which the high substitution by galactose obviously establishes so much steric hindrance that it dissolves easily in water and prevents attack for some time from degrading enzymes.

As has been pointed out, galactomannans show certain properties:

(1)   As with many polysaccharides, they are rich in hydroxyl groups. This enables them to bind and take up water.

(2)   They are rich in cis-OH groups which allow aggregation from chain to chain via hydrogen bonds - so that hydration becomes more complicated if interchain crosslinking can take place.

By substitution with galactose, nature establishes steric hindrance between the molecules and thus enhances (colloidal) water solubility.

Chemical Derivation of Galactomannans

This effect can also be achieved by chemical substitution. This brings all accessible galactomannans into a cold water soluble form, whether or not they were originally soluble. Some substitution reactions on galactomannans have gained technical interest and have therefore grown to commercial relevance.

Nonionic substitution, for example, takes place if the galactomannan is reacted with chlorohydrin, ethylene oxide or propylene oxide in alkaline medium. From guar, tara, cassia or locust bean gums the reactions yield clear collodial water soluble products.

Substitution with anionic substituents such as   etherification with sodium-carboxymethyl or -carboxyethyl-groups adds additional hydration-effects due to charged groups to those caused by steric hindrance. Further anionic substitution is done by phosphorylation, which also leads to colloidal, clear, soluble, highly viscous phosphoric acid esters[12,18].

Substitution with cationic reaction partners leads to cationic galactomannans - especially by reacting ethylene-imine or trimethyl-glycidyl-ammonium with the different galactomannans.

In many cases the relatively high molecular weight of most galactomannans is advantageous but sometimes a viscous solution needs more body, more dry substance to achieve better application properties etc. In these cases some degradation of the high molecular weight is necessary. This can be carried out with all galactomannans by acid hydrolysis or oxidation with peroxides or even air. Such degraded galactomannans show interesting application properties in food-processing, paper-production or textile printing and sizing.

Technical Application of Galactomannans

The world figures of galactomannan consumption differ quite considerably according to different sources. Our company assumes that around 90 - 100 thousand tons are consumed per year. The biggest consumption being that of guar gum with

70 - 80 thousand tons, followed by locust bean gum with 12 - 14 thousand tons. The leading consumer is the food industry.

Reviewing the technical applications, attention should be drawn to the chemical behaviour of the different galactomannans. There are applications which benefit from the excellent viscosity formation of some galactomannans or their derivatives and there are also applications which benefit from water absorption or from the formation of hydrogen bonds as well as gel formation.

<u>Viscosity formation</u>. Some galactomannans, especially guar, but also cassia and locust bean gum derivatives, develop very high viscosities in aqueous solution. In the textile industry they are used to thicken the dyebaths in printing and dyeing of fibers, fabrics and carpets. The gums control the flow characteristics of the dye formulations, so that sharp, bright patterns can be achieved. For different types of cloth and dyestuff, different types of thickener are used - mostly combinations. Table 2 shows which type of galactomannan - derivatives are used for the different groups of dyestuffs and textiles.

<u>Table 2</u>   Application of galactomannans and their derivatives in textile production

<u>Basis-Galactomannan</u>	<u>Chemical modification</u>	<u>Preferred field of application</u>
G U A R A N Galactomannan - 1.2	unchanged	carpet - printing /-dyeing acid-, metalcomplex - dyestuffs
	depolymerized	carpet - printing - /-dyeing acid-, metalcomplex - dyestuffs
		printing of: cotton / viscose vat-, direct-, fastsalt-dyestuffs
		polyester dispers - dyestuffs
		polyamide acid-, metalcomplex -dyestuffs
		polyacryl catonic - dyestuffs
	hydroxyethylated	carpet - printing /-dyeing acid-, metalcomplex - dyestuffs
		cotton africa -print with fastsalt-dyestuffs
		cotton /viscose vat - dyestuffs
		polyester dispers - dyestuffs
		polyamide acid-, metalcomplex -dyestuffs

		polyacryl cationic - dyestuffs
	hydroxypropylated -CH$_2$-CHCH$_3$ - OH	carpet - printing /-dyeing acid-, metalcomplex -dyestuffs
	additionally depolymerized	textile sizing
	carboxymethylated -CH$_2$ - COO$^-$Na$^+$	cotton - printing carpet acid-, metalcomplex -dyestuffs
		cotton vat-, reactive- dyestuffs (restr.).
C A R U B I N Galactomannan - 1.4	hydroxyethylated -CH$_2$ - CH$_2$ - OH	carpet - printing /-dyeing acid-, metalcomplex -dyestuffs
		cotton africa -print with fastsalt-dyestuffs
		polyester dispers - dyestuffs
		polyamide cationic - dyestuffs
		wool / silk acid-, metalcomplex -dyestuffs
	carboxymethylated -CH$_2$ - COO$^-$Na$^+$	cotton / viscose vat - dyestuffs
		wool / silk acid - dyestuffs
C A S S I A Galactomannan-1.5	hydroxypropylated -CH$_2$CHCH$_3$OH additionally depolymerized	textile sizing
	carboxymethylated -CH$_2$ - COO$^-$Na$^+$	carpet - printing /-dyeing acid-, metalcomplex -dyestuffs
	additionally depolymerized	cotton / viscose vat - dyestuffs
		wool acid - dyestuffs

In the field of explosives, guar gum and derivatives are used to thicken nitrate salts solutions, which are the basic components of slurry explosive formulations. These compositions are safer to use and can be formulated to demand as a viscous liquid or even as gels.

It is well known that the productivity of oil and gas wells can be increased by cracking and opening up the oil or gas bearing zones with hydraulic pressure. Hydroxypropylguar-solutions and other derivatives are used in this process which is known as hydraulic fracturing. A highly viscous liquid, carrying sand, is pumped under very high pressure into the fractured rock. The sand keeps the fracture open, when the hydraulic pressure is released. Oil and gas are then recovered at an increased rate.

In this aspect another property of galactomannans is also of advantage: the ability to gel with elements like boron or transition metals in the form of their salts, which can be used to block or tighten wells. Since this gelling reaction is reversible with change of pH, it is used to control fluid loss in wells.

Gel Formation

As has been shown, galactomannans are able to form gels with certain metal salts. This effect is used in the field of textiles to print vat-dyestuffs in two-phases, which yields bright and sharp prints[11]. In such gels the crosslinking reaction has been found to take place mainly by reaction with cis-OH-groups of the galactose-units.

On the other hand galactomannans effect carrageenin- and agar-gels[9,10] which according to Dea and Morrison[2] diminish from one galactomannan to another in the direction of increasing content of D-galactose.

Thus it can be assumed that the mannose residues of the main chain are mainly involved. Therefore, only galactomannans with little or no steric hindrance by galactose groups may show this effect. When combining carrageenin with different galactomannans no effect is found with galactomannan-1,1, very little with guar, little with tara-galactomannan-1,3, whereas good effects occur with carube-galactomannan-1,4 which are even exceeded by cassia-galactomannan-1,5[17]. Such gels have interesting properties: a 0.3% kappa-carrageenin / 0.3% cassia-gum aqueous solution after boiling and settling to room temperature reaches gel-strength rates of 80 g with the FIRA Jelly Tester whilst similar gels with locust bean gum show less and with guar even much less strength. Such properties will be useful in food and petfood but also in technical applications, where gels are used to bear and bind odours, as in solid airfresheners for example. Transport of solid powders like coal or ores suspended in such gels through pipelines could also be possible as these gels have an excellent suspending force, which can be destroyed immediately when reaching the final point of destination by heating to only about 50°C.

<u>Applications benefiting from the formation of hydrogen bonds</u>. As pointed out earlier, easy formation of hydrogen bonds is one of the characteristics of galactomannans. This is used widely in the paper-industry (1980: 7.000 tons) where guar has replaced locust bean as a wet end additive. The pulping process serves to remove lignin and a large part of hemicelluloses which are normally present in the wood. Galactomannans replace the natural hemicelluloses in paper bonding. It is generally agreed that the hydrogen-bonding effect is one of the major factors in fibre-fibre bonding since the galactomannans absorb onto the hydrated cellulose fibres. Machine speed is increased by using galactomannans and the retention of fine fibres during the process is also increased.

CM-derivatives of cassia gum also show excellent hydrogen-bonding effects - especially when used as thickeners in paper sizing. Thus light-weight-papers can be produced with excellent tightness, so that printing inks cannot strike through.

In the mining industry galactomannans are used as chemical flotation agents rather than as thickeners. The ability to form hydrogen bonds is used here to absorb the gum onto hydrated mineral surfaces. In flotation, the galactomannan functions as depressant to block the absorption of other reagents onto the surfaces of talc and other gangues which are mined along with the valuable minerals. Also, by linking for example the montmorillonite particles together, this effect depresses the gangues so they can settle during the flotation operation and the ore particles are free to absorb the collector-chemicals and can then be floated.

Water absorption.   To protect water sensitive goods, they may be covered with galactomannan powder and then covered again. If the outer cover breaks and water should enter, then the galactomannan swells and prevents water from further penetration. Dry guar is used as a cover substance to prevent water from entering into cartridges filled with explosive powders, cables and so on.

Crosslinked galactomannan-derivatives lose their ability to dissolve. Dependent on the grade and kind of substitution they do not lose their ability to absorb water. These substances can be used in competition with synthetic polymers as absorbents for hygiene articles.

Obviously, the applications of galactomannans depend more or less on their ability to form hydrogen bonds, a property which is counterplayed by the steric effect of galactose side-groups, which help to achieve increased solubility. The most interesting galactomannans are found on the extremes either with very high or very low substitution; the latter have been found in cassia gum.

Cassia Gum.   The genus Cassia L. is the largest genus in the subfamily Caesalpinioideae of the Leguminosae and is probably one of the twenty five largest genera of dicotyledons. Irwin and Barneby[15] have defined three genera, i.e. Cassia, Senna and Chamaecrista.

Cassia tora, which was originally synonymous with Cassia obtusifolia ranges therefore subsequently under Senna tora and Senna obtusifolia, which two species are now considered to be distinct, although their chemical composition is almost identical and their gross morphology very similar. According to Kew Index Cassia tora has 12 synonyms. Taxonomic investigations show almost no difference between the synonymic plants.

The plant can be described as annual herb or subshrub, up to 1 m in height. The plant's characteristics are: stem angular, glabrous; leaves 7 - 15 cm long; leaflets 3 pairs opposite; bright-yellow flowers usually in subsessile pairs in the axil of the leaves; pods 20 - 25 x 0.5 cm; seeds 30 - 35 rhombohedral, 5 mm long[14, 15, 16].

Cassia tora/obtusifolia grows in the tropical and subtropical belts around the world in an almost circumtropical dispersal. It seems to be reliably established that Cassia obtusifolia was native to South America and has been transported as a weed to tropical areas of Africa, Asia and the Pacific[15]. Similarly, it seems highly probable that Cassia tora was native somewhere in the Asian-Pacific region and has been transported as a weed to Africa and Australia. They are interseedling and thus found in many regions of India[14].

The seeds of Cassia tora/obtusifolia are the basic raw material for the production of the galactomannan-1,5 (DIAGUM CS). The harvesting of the seeds is carried out in the country of origin after maturation of the plant, which in India is from November to the end of January. The pods are harvested and the grains are separated by threshing. In a first separation step stones, dust and foreign matters are removed. The whole grains when heated to 180° C become fragile. When crushed in refining equipment, two different fractions are obtained: on the one side, the germ and the husk and on the other side, the endosperm. The endosperm is cleaned further and milled to different mesh sizes (100, 200, 300) depending on the intended application.

The 200 mesh powder serves as a gelling aid, together with carrageenin, xanthan etc. as mentioned before. Other mesh sizes are substituted by carboxymethyl- or hydroxy-propyl groups for the production of textile- and paper-auxiliaries.

Analytically we have found no significant differences between the seeds of the synonymic plants, neither in the carbohydrate composition nor in fat or protein analysis. The byproducts have been analyzed and compared by Poethke[13] who also found Cassia tora and Cassia obtusifolia to be identical.

Of course, we do not find one uniform galactomannan in the endosperm but water soluble fractions and a hot water insoluble fraction which differ in their galactose / mannose ratio from 1 : 4,9 to 1 : 7. The hot-water-insoluble fractions consist of almost unsubstituted polymannan.

All the fractions show properties which are to be expected from a galactomannan with relatively large areas of sterically unhindered cis-OH-groups. Therefore, as already mentioned, the technical applications benefit from the excellent film forming properties usable in textile sizing agents, paper sizing agents and textile printing thickeners. The ability to form synergistic gels with carrageenin, xanthan, polyacrylates and several other polymers, should also be mentioned.

In reviewing the group of galactomannans, it can be stated that the extreme outsiders like galactomannan-1,5 - with low steric hindrance - or galactomannan-1,1 with full steric hindrance - are offering the most interesting prospects for research and application.

3 REFERENCES

1. P.M. Dey, Adv. Carboh. Chem. Biochem., 1978, 35, 347.
2. Jain C.M. Dea, A. Morrison, Adv. Carboh. Chem. Biochem., 1975, 31, 269 .
3. H.L. Tookey, Q. Jones, Ec. Botany, 1965, 19, 165.
4. V.P. Kapoor, M.J.H. Farooqi, L.D. Kapoor, Indian Forester, 1980, 106, 810.
5. S.K. Katiyar, G.S. Niranjan, J. Indian Chem. Soc., 1981, 58, 98.
6. M.J.H. Farooqi, V.P. Kapoor, P.S.H. Khan, Res. Industry, 1985, 30, 144.
7. P.C. Gupta, S. Mukherjee, Indian J. Chem., 1973, 5, 505 f.
8. H. Neukom, P.A. Hui, Tappi, 1964, 47, 39.
9. G.L. Baker, J.W. Carrow, C.W. Woodmansee, Food Ind., 1949, 21, 617.
10. H. Deuel, G. Huber, J. Solms, Experientia, 1950, 6, 138.

11. W. Jülicher, H. Appelt, DE-PS 896795, Diamalt, 1944.
12. USP 4162925, EP 0030443 A. Meyhall
13. W. Poethke, Rao D. Anand, K.D. Löscher, Pharm. Zentralh., 1986, 107, 571.
14. M.N. Bhandara, Flora of the Indian Desert, Scient. publ., 1978, 142.
15. H.S. Irwin, R.C. Barneby, Mem. New York Bot. Gard., 1982, 35, 252.
16. B.R. Randell, J. Adelaide Bot. Gard., 1988, 11 (1), 19.
17. F. Bayerlein, M. Kuhn, M. Maton, EP 0139913B2, Diamalt, 1984.
18. F. Bayerlein, U. Beck, N. Keramaris, M. Kuhn, N. Kottmair, M. Maton, EP 0146911, Diamalt.

EDIBLE FIBER FROM BARLEY AND OATS BY WET-MILLING

A. Lehmussaari[1] and M.G. Lindley[2]

[1] ALKO LTD., 05200 RAJAMÄKI, FINLAND
[2] LINTECH; BOX 62, READING, RG6 2BX, UK

1 INTRODUCTION

There are growing consumer pressures on food manufacturers, particularly in the United States, to reduce the number and amounts of chemical additives and chemically modified additives used in foods. As part of this movement, product formulation opportunities which avoid using the term "modified starch" on product labels are increasingly being sought.

The food industry has long had a number of native food starches available for use in its product development, including starch from corn, wheat, potato, tapioca, and rice. These starches, while providing useful functionalities, clearly are unable to offer the full range of stabilizing and thickening characteristics delivered by modified starches. Therefore, in an attempt to meet the demands of the food industry for native, unmodified starches of appropriate functionalities, Alko has developed industrial scale wet milling processes to produce high quality starches from oats and barley. These starches have recently been made available to industry which is proceeding with its evaluation of their value. Having extracted the starch from these grains by wet milling, further fractionation may then be applied to obtain protein and fiber enriched fractions. Therefore, oat and barley fiber enriched ingredients are also now being produced by Alko in commercial quantities.

In this paper, the process whereby these fibers are produced is described. Details of their compositions and the influence these compositions have on functional properties are presented. The effects of the barley fiber on various physiological parameters and of this oat fiber on serum lipids in hypercholesterolemic subjects have also been studied, and the results of these physiological studies are described. Finally, results of a range of food product applications trials are reviewed.

2 BARLEY AND OAT PROCESSING

Dry and wet milling methods are both utilized for the processing of oats and barley. In principle, the basic processes employed to process both grains are very similar, as shown in Figure 1. Due to the very different nature of these grains, the conditions used for their processing also differ[1-3].

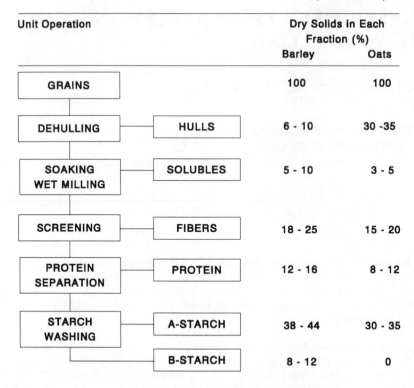

Unit Operation		Dry Solids in Each Fraction (%)	
		Barley	Oats
GRAINS		100	100
DEHULLING	HULLS	6 - 10	30 -35
SOAKING WET MILLING	SOLUBLES	5 - 10	3 - 5
SCREENING	FIBERS	18 - 25	15 - 20
PROTEIN SEPARATION	PROTEIN	12 - 16	8 - 12
STARCH WASHING	A-STARCH	38 - 44	30 - 35
	B-STARCH	8 - 12	0

<u>Figure 1</u>   Basic process scheme for combined dry and wet milling of oats and barley.
Numbers refer to typical mass balances in the process given in percent dry
substance

Hulling of barley and oats is carried out using a hammer mill, followed by aspiration and sieving. Production of high quality edible fiber from oats uses the preferred procedure whereby groats are first made to guarantee that no hulls will be present in the final fiber fraction.

The coarse flour is then mixed with water and soaked. Cellulases and β-glucanase are added to the barley to soften the cell structures and reduce viscosity in downstream processing. This treatment hydrolyses the β-glucan, thus leaving a barley fiber product which contains very little soluble fiber. Oats may be processed following a very short soaking period, thus retaining most of the β-glucan still embedded within the cell structures. The resulting oat fiber thus contains ca.15% β-glucan.

Protein and starch are fractionated by centrifugal separation. Most of the lipids end up in the protein fraction, thus restricting the shelf life. For this reason, the protein fraction is used principally in animal feedstuffs.

Starch is then washed by hydrocyclone technique. Barley contains a small quantity (4-8%) of small granule B-starch which may either be combined into the protein fraction, going to animal feed, or hydrolysed to sugars to be fermented to alcohol.

Barley starch is very similar to other cereal starches (e.g., wheat and corn) and is mainly used by the paper and food industries. Some starch is also used in brewing, glucose and high fructose syrups production. Oat starch granules are very small, with an average diameter of 4-6 microns. All oat starch granules are below 10 microns.

3 PRODUCT COMPOSITIONS AND PHYSICAL PROPERTIES

The analytical compositions of these fiber products have been measured and the results are summarized in Tables 1 and 2.

Table 1 Typical analysis of barley and oat fibers

Component	Composition (%)	
	Barley fiber	Oat fiber
Total dietary fiber[1]	70	35
of which soluble	1	14
of which insoluble	69	25
Protein	11	25
Starch	11	25
Fat	6	11
Minerals	3	2

[1] AOAC Method

Table 2 Typical dietary fiber composition of barley and oat fibers

Component	Composition (%)	
	Barley fiber	Oat fiber
Hemicellulose	68	85
Cellulose	23	10
Lignin	5	4
Pectin	1	1
Rhamnose	0.3	0.2
Arabinose	23.3	12.4
Xylose	39.5	20.2
Mannose	2.2	1.8
Galactose	2.7	1.8
Glucose	28.5	59.4
Uronic acids	3.5	4.2

The functional properties of these fibers have also been determined, as follows. The rate of and capacity to take up water have been measured in a Baumann apparatus. Water up-take, followed during a 30 minute period, shows that barley fiber has a higher water up-take capacity than oat fiber. This is not surprising given its high insoluble dietary fiber content. In contrast, oat fiber has much higher water-holding and water-binding capacities than barley fiber, results which reflect the high soluble fiber content of the oat fiber. Probably for the same reason, oat fiber fat-binding capacity is much lower than that of barley fiber (Table 3).

Table 3   Functional properties of barley and oat fibers

Property	Barley fiber	Oat fiber
	ml/g	ml/g
Water up-take capacity		
1 minute	3.6	1.3
15 minutes	4.0	2.6
30 minutes		2.8
Water holding capacity		
18 hours	5.4	9.3
Water binding capacity		
1 hour	1.7	3.1
3 hours		4.9
18 hours		6.2
20 °C	1.7	3.1
80 °C	2.1	8.3
Fat holding capacity	0.8	0.3

Taken together, these compositional results and functional properties suggest that these fibers will exhibit quite different physiological characteristics. They also suggest that potential food product applications are unlikely to show significant overlap.

4   PHYSIOLOGICAL PROPERTIES

The United States Surgeon General's Report on Nutrition and Health states that the prevalence of diseases such as coronary heart disease, cancer and obesity may be associated with a dietary pattern in which foods high in fat are consumed at the expense of foods high in complex carbohydrates, including dietary fiber. Clearly the incidence of these diseases is multi-factorial, and the specific association between dietary fiber and disease prevention may be difficult to prove, but there are generally recognized benefits of consuming foods high in dietary fiber.

It is thought that dietary fiber, in one or more of its entities, may be a positive factor in maintaining health by promoting increased fecal moisture, volatile fatty acids and bulk, and by decreasing fecal transit time. These help to avoid constipation and may help lower blood cholesterol and triglyceride levels. Therefore, with some justification, dietary fiber is becoming associated in the public mind with the prevention or treatment of a variety of diseases or disease conditions, including diverticulitis, hemor-rhoids, ulcerative colitis, diabetes, cancer and coronary heart disease.

Given the composition of these Alko fibers, and the functional properties they ex-hibit, both have been evaluated for their impact on selected clinical conditions. The in-fluence of barley fiber on fecal weights and transit times has been measured[4], as has its potential to balance blood glucose levels in diabetic patients[5]. Oat fiber has been as-sessed for its ability to influence the cholesterol levels of hypercholesterolemic pa-tients[6].

Barley Fiber

Controlled studies designed to measure the ability of barley fiber to increase fecal weight and to reduce fecal transit times have been completed, as well as studies to measure the effects on blood glucose levels. Details of these studies were as follows.

Influence of Barley Fiber on Fecal Weight.   According to the literature, coarsely ground wheat bran is the most effective fiber source to increase fecal weight. Wheat bran reportedly increases fecal weight by 80-100%. This literature value compares with other literature values for oat bran (15%), apple fiber (40%), carrot fiber (60%), and cellulose (75%). Therefore, this barley fiber, increasing fecal weight by an average 76%, may be considered to be a very effective aid to increasing fecal weights (Table 4).

Table 4   Summary of protocol and findings of a study to measure the influence of barley fiber on fecal weight[4]

Subjects	n = 14 (female; ages 22-26)
Duration	control 12 days + test 12 days
Dietary fiber consumption	
control period	15.7g/day
test period	34.6g/day
Increase in dietary fiber consumption	18.9g/day
Fecal weight	
control period	106g/day
test period	187g/day
Increase in fecal weight (c.f. control)	76%

Influence of Barley Fiber on Fecal Transit Time.   Fecal transit times were mark-edly reduced as a consequence of consuming these amounts of barley fiber, results which might be expected given the influence of the fiber on fecal weights (Table 5).

Fecal transit times were markedly reduced as a consequence of consuming these amounts of barley fiber, results which might be expected given the influence of the fi-ber on fecal weights.

Influence of Barley Fiber on Blood Glucose Levels.   Diabetic subjects consumed 30g barley fiber on a daily basis. Once each week, patients ate a test meal, after which

Table 5   Summary of protocol and findings of a study to measure the impact of barley fiber on fecal transit time[4]

Subjects	n = 40, all constipated, ages 20-81
Duration of test	3 weeks
Duration of symptoms	1-50 years
Dietary fiber consumption	
control period	14.9 g/day
test period	21.7g/day
Increase in dietary fiber consumption	5-7g/day

Frequency	Control period	Test period
Daily	0 persons	18 persons
Every 1-2 days	8 persons	19 persons
Every 3-4 days	29 persons	3 persons
Less frequently	3 persons	0 persons

the serum glucose levels were measured. A control period without barley fiber pre-
ceded the test period. Results are summarized in Table 6[5].

Table 6   Influence of barley fiber on the increase of blood glucose levels[5]

Subjects	n = 12 (diabetics aged 57-83 years)
Duration of test	5 week control period
followed by	5 week test period
Consumption of dietary fiber	30g/day

Increase of blood glucose levels following test meals

Time after meal (min)	Control period	Test period
	(mg glucose/100ml blood)	
30	55.4	51.6
60	85.6	58.1*
90	75.0	46.0*
120	53.3	19.6**

$* p < 0.05; ** p < 0.02$

From the data in Table 6 it is clear that increases in serum glucose levels following a test meal were reduced significantly by consumption of barley fiber during a test
period as compared to a control period with no added barley fiber.

Oat Fiber

In several studies soluble dietary fibers have been shown to lower serum total
cholesterol and low-density lipoprotein cholesterol levels. The exact mechanism by

which cholesterol reduction occurs is not known, but it seems likely that the soluble β-glucans in oat and bran are the active components.

A controlled study on the hypocholesterolemic properties of Alko's oat fiber product has been carried out and the detailed findings reported[4]. A short resume of the protocol is presented in Table 7.

Table 7   Baseline characteristics and fiber/bran consumption

		Oat bran group	Wheat bran group
Sex	Men	n = 10	n = 10
	Women	n = 10	n = 6
Age (years)		50+/-6	45+/-9
BMI (kg/m^2)			
0 weeks		26.3+/-3.3	26.7+/-2.5
8 weeks		26.3+/-3.3	26.6+/-2.6
Fiber consumption (g/day)			
0 weeks		22.7+/-8.1	24.4+/-5.5
4 weeks		21.5+/-7.9	21.8+/-5.9
8 weeks		20.9+/-7.6	19.0+/-6.4

At the end of 4 weeks, no changes in the total serum cholesterol and low density lipoprotein cholesterol of the wheat bran group were found, but in the oat bran group, both parameters had decreased significantly. After 8 weeks, however, the values for both parameters in both groups were not significantly different from those measured at baseline. Although these results suggest that the hypocholesterolemic effect of oat bran may diminish with time, long term studies are necessary to evaluate the value of oat bran products as an adjunct therapy for hypocholesterolemia.

5   FOOD PRODUCT APPLICATIONS

As suggested in the Introduction to this chapter, the composition and functional characteristics of these barley and oat fibers show few similarities. This suggests their potential food applications are unlikely to overlap to any significant degree.

Barley fiber is essentially inert, being largely comprised of insoluble dietary fiber. This suggests potential applications in foods where fiber fortification is required. In contrast, the oat fiber, while high in dietary fiber relative to other oat bran products, typically contains only c.20% insoluble fiber and is high in soluble β-glucan. It also contains substantial quantities of protein, oat starch and fat. Thus, it is a more functional fiber than barley fiber and as such will more readily influence the structure of foods in which it is incorporated.

Barley Fiber

The product evaluation programme to assess the potential of barley fiber has concentrated on some applications in baked products. Barley fiber is able to function as a fiber source in bread doughs and in biscuit products where it may also successfully replace some flour in the formulation. For example, barley fiber has been assessed in the biscuit formulation (Table 8).

Table 8   Formulations to evaluate barley fiber in biscuits

Ingredient	Standard	Barley fiber biscuit
	(Parts by weight)	
Biscuit flour	49.4	41.4
Barley fiber	0.0	8.0
Wholemeal flour	6.1	6.1
White shortening	17.2	17.2
Sugar	13.3	13.3
Sugar syrup (83oBrix)	4.0	4.0
Salt	0.6	0.6
Sodium bicarbonate	1.0	1.0
Tartaric acid	0.4	0.4
Water	8.0	8.0

Although there were differences between biscuits made with these formulations, the barley fiber biscuit had no bitterness or off-tastes. Texture, crumb structure and crunch were judged to be 'good', although there was some surface cracking. The principal difference observed during manufacture was that for effective dough mixing, more energy was required by the barley formulation due to its water holding capacity. Despite this, very acceptable products were produced.

Other applications trials have simply focussed on evaluating barley fiber as a fiber source added to cereal products. Its quite neutral taste and color make it compatible with muesli cereals and muesli cereal bar products.

Oat Fiber

Alko's oat fiber has been evaluated more broadly as an ingredient in extruded snacks, extruded breakfast cereals, muffins and baked cereal bars.

Two product examples which illustrate the versatility of the oat fiber are extruded breakfast cereals and muffins. Both these products have traditionally been difficult to produce in easy eating, highly palatable forms, but use of Alko's oat fiber, with its high soluble fiber element, allows the preparation of high quality forms of such products.

Examples of breakfast cereals have been prepared by extruding the formulation presented in Table 9 using a twin-screw extruder.

Table 9   Extruded cereal formulation containing oat fiber

Ingredient	Amount (%)
Alko oat fiber	40.0
Rice starch	20.0
Corn flour	19.0
Sucrose	10.0
Oat starch	5.0
Malt flour	5.0
Salt	1.0

Products prepared according to this formulation had light, crisp, even textures. On adding milk, as in a normal eating situation, the cereal retained its crisp, easy to chew texture.

It might be expected that a product such as this extruded cereal, which has no water in the formulation, and has little opportunity to fully hydrate during production, would successfully accommodate this high level of fiber. However, we have also been able to prepare excellent quality baked products such as muffins, containing high levels of oat fiber, whose texture and eating quality were judged as excellent (Table 10).

Table 10   High fiber muffin formulation incorporating oat fiber

Ingredient	Amount (%)
Flour	20.7
Milk	18.1
Sultanas	14.3
Alko Oat fiber	12.6
Whole egg	10.9
Sucrose	10.5
Butter	5.9
Fructose	4.5
Vegetable fat	1.1
Baking powder	1.0
Salt	0.4

Muffins prepared according to this recipe had the same light texture, pleasant mouthfeel and volume as the controls, thus illustrating the versatility of this oat fiber.

6   CONCLUSIONS

Alko's barley and oat fiber ingredients have been produced by a unique process as by-products of barley and oat starch fractionation schemes. Both fibers have been shown to induce beneficial physiological effects under controlled clinical testing. The barley fiber, being essentially inert, can be used in foods, particularly baked products, where fiber fortification is required. Oat fiber is a more functional ingredient, contributing positive benefits to baked goods, extruded snack products, breakfast cereals and even to processed meats. Both ingredients should find ready acceptance by European and North American food processors.

7 REFERENCES

1. European Patent Application: EP 0 381 872.

2. U.S. Patent 4,957,565.

3. U.S. Patent 5,106,640.

4. E. Wisker, G. Peschel and W. Feldheim, Akt. Ernähr., 1988, 2, 52.

5. A. Werthmann, 'Der Einfluss von Ballaststoffen aus Zuckerrüben und Gerste auf den Kohlenhydrat- und Lipidstoffwechsel', Master Thesis, University of Kiel, 1988.

6. M.I.J. Uusitupa, E. Ruuskanen, E. Mäkinen, J. Laitinen, E. Toskala, K. Kervinen and Y.A. Kesäniemi, 'Topics in Dietary Fiber Research', Paper presented at the COST 92 Programme Symposium Rome, May, 1992.

Chemical and Enzymic Conversion

SOLUTION PROPERTIES OF PLANT POLYSACCHARIDES AS A FUNCTION OF THEIR CHEMICAL STRUCTURE

W. Burchard

INSTITUTE OF MACROMOLECULAR CHEMISTRY, UNIVERSITY OF FREIBURG, STEFAN-MEIER-STRASSE 31, W-7800 FREIBURG, GERMANY

ABSTRACT

In contrast to common, non-polar synthetic polymers biological macromolecules show strikingly different behaviour in solution. These differences arise mainly from the complex chemical structure of the repeating units, their polar side groups and the occurrence of branching. The large number of different possible glycosidic linkages stimulates preformed suprastructures for the macromolecule. For instance the β-(1,4) glucosidic bond in cellulose causes a stretched chain conformation which induces fibre formation whereas the α-(1,4) bond favours helix formation and the well known complexing capability of amylose. The OH- are responsible for hydrogen bonding, which is the main cause for the high tendency of association or aggregation.

For the study of structural and conformational properties three types of methods can be efficiently employed. These are: viscometry and oscillatory rheology, static and dynamic light scattering and a combination of size exclusion chromatography with on-line light scattering and viscometry. Only the possibilities of static and dynamic light scattering are discussed with examples of cellulose derivatives, hemicelluloses and pectin.

With the exception of only a few examples cellulose and its derivatives do not form molecularly dispersed solutions. Measurements at different molar masses reveal an unexpected weak increase of the radius of gyration, which is interpreted by a lateral (side-by-side) aggregation of chains. The conclusion could be confirmed by dynamic light scattering according to which an increase in segment density occurs as the molar mass increases. These findings led to the model of a fringed micelle.

Pectins are available as industrial products, they are not well defined, contain fractions from other cell wall polysaccharides and have made a clear clarification of the structure infeasible. Progress has been achieved by better isolation techniques in combination with a comprehensive enzymic treatment. Light scattering measurements from an arabinogalactan obtained from apple juice by this technique reveals a much larger, highly branched structure than was expected from GPC.

1   INTRODUCTION

In the past the research on polysaccharides was mainly concerned with the determination of the sugar composition, with methods for chemical and enzymic modification or with the structure and function of such enzymes. Interest was focused mainly on the local chemical structure, since this often gives sufficient information on the specific interactions between the enzyme and the polysaccharide or the polysaccharide and a specific surface. Recently attention is paid to the question how the abundantly available polysaccharides can be economically used as a source for new, biodegradable materials. The economical use of such materials and their optimized application requires the study of behaviour of these macromolecules in three dimensional space.

In the field of synthetic polymers it is known that this three dimensional polymeric structure gives rise to the desired properties of plastic materials. It has been shown by de Gennes[1] that with flexible non-polar chains the chemical nature plays only a subordinate role and that *topological* properties govern the whole behaviour when the chain is sufficiently long. Typical topological quantities are the chain length, the bond length and bond angles, branching points and the chain sections between two branching points, the volume which is spanned by a macromolecule and the corresponding radius of gyration. Universal laws have been proposed which immensely facilitated the study of synthetic polymers. One may wonder here, whether such topological considerations may also be applicable to the biological materials and, in particular, to polysaccharides.

The large number of different linkages in polysaccharides suggests expectations that such a consideration would give a good guide for a better understanding of the complex polysaccharide behaviour in solution. Unfortunately the situation is far more complex since the many highly polar OH-, anionic or cationic charges, or N-acetyl amino groups, cause strong interactions. The first impression is that the universality laws as formulated by de Gennes no longer hold, since the special chemical structure now plays the dominant role. Nevertheless, a closer look at the influence of the special linkages and their influence on the arrangement of the chains in space appears to be of interest and will be considered with a few examples.

In a second step the large number of possibilities for association and aggregation will be examined and this will lead to the unexpected observation that a classification into various groups can be made. For chains which are strictly regularly built up by unique repeating units this outcome may be understandable, and the chemical structure gives rise to well defined suprastructures as found with bacterial polysaccharides. Such a strict regularity is not observed with most plant polysaccharides, and we have to wonder why this is so. This question will be discussed with special examples.

2   CONSEQUENCES OF SPECIFIC LINKAGES

Let us for a while forget the many OH-groups, or other polar groups, and consider the sole effect of the linkages between the various sugars. The corresponding lowest energy conformation can nowadays be found by techniques which were outlined by Flory in 1969[2] and these techniques have been very fruitful for the estimation of possible crystalline structures. These procedures are time consuming and require special experience with computer modelling. A fairly good insight can be gained, however, by joining the sugars in the simple manner as demonstrated in Figures 1-3 in which all

OH-groups are not shown. The even further simplified skeletal structure is shown underneath.

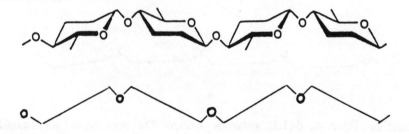

<u>Figure 1a</u>   Effect of β-(1,4) linkage. A stretched zig-zag chain is obtained; the group in C6 position points alternately in opposite directions. Example: cellulose

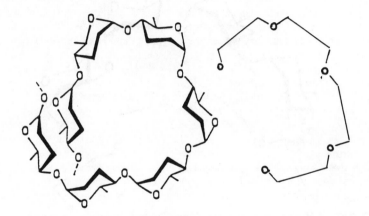

<u>Figure 1b</u>   Effect of α-(1,4) glucosidic linkage. A wobbled helix is preformed; the group in C6 position points in the same direction. Example: amylose

In Figures 1 and 2 the effect of equatorial and axial linkages is compared for the (1,4)- and (1,3)-glucosidic bonds. The effect of the β-(1,4) linkage is well known: each anhydroglucose unit appears rotated by 180 degrees, such that the C6 group of a dimer is pointing towards opposite directions of the chain. The chain corresponds to a flat zig-zag chain (or a two-fold helix) and has the effect of a preformed fibrous conformation. The α-(1,4) linkage , on the other hand, causes a preformed wobbled helix with 6 to 8 residues per turn. Both conformations, the flat β-(1,4) and α-(1,4) helical chains will not be observed in solution, since the rotation angles are not fixed but possess a certain freedom. Hence additional forces are needed to stabilize these conformations. In cellulose there are special H-bonds which stabilize the conformation intramolecularly and simultaneously cause a crystallization via intermolecular H-bonds: the material becomes completely insoluble. There exist also H-bonds between maltosyl units but the effect of these H-bonds is not sufficient for the helix stabilization. Hence other complexing agents are needed for such a stabilization.

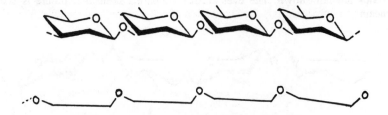

<u>Figure 2a</u>   Effect of β-(1,3) glucosidic linkage. The units appear translationally shifted. The groups in C6 position point in the same direction. For steric reasons twisting occurs

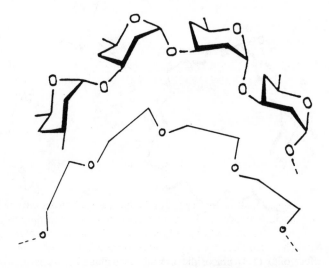

<u>Figure 2b</u>   Effect of the α-(1,3) glucosidic linkage. A helix similar to that of amylose is stimulated. Example: bacterial polysaccharides

The effect of the (1,3)-linkages (Figure 2a and b) is similar to that of the (1,4)-linkage with an important difference: in the β-(1,3) bond the anhydroglucose units are not rotated. One obtains a conformation where the units appear to be simply translationally shifted along a rodlike line. The C6 groups point here in the same direction. For steric reasons, i.e. a better space filling, the chain tends to become twisted. Triple helices are often found which can be explained by special H-bond formation between the three chains. The suspected helix formation of the α-(1,3) chain is observed with bacterial polysaccharides, but to my knowledge a pure α-(1,3) homopolymer has not been observed.

The effect of a β-(1,2) linkage is demonstrated in Figure 3a. A sharp kink is introduced in a chain that may be rather regular. Hence the crystallization as observed with cellulose becomes strongly hampered. The effect of the α-(1,2) linkage on the macromolecular conformation is similar, but the local structure is, of course, very different and certainly important for recognition of cell surfaces and the specific antigen-antibody complex formation.

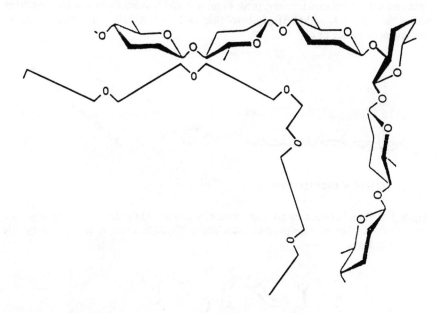

<u>Figure 3a</u> A β-(1,2) linkage introduces a sharp kink

As a last example the structure of an α-(1,4) galactan section is shown in Figure 3b. Again a zig-zag chain is expected similar to that of cellulose, but the chain is not as much extended and more narrowly folded. A stronger hindrance of rotation, resulting in a pronounced chain stiffness, may be deduced from this fact but remains questionable for reasons, which are discussed later in this contribution.

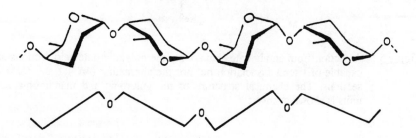

<u>Figure 3b</u> Effect of the α-(1,4) galactosidic linkage. A narrower zig-zag chain as in cellulose results

3   ROLE OF IRREGULARITIES IN MACROMOLECULAR STRUCTURE

A fully regular homopolymeric structure is rare in the realm of plant polysaccharides. Cellulose and amylose are probably the only examples of such regular structures. In most cases the regularity is interrupted. Figures 4 and 5 show two examples. Another one is the $[\beta(1,4)]_x$ -$[\beta$-$(1,3)]_y$ glucan that is found in cell walls of the starchy endosperm of barley or oat[3].

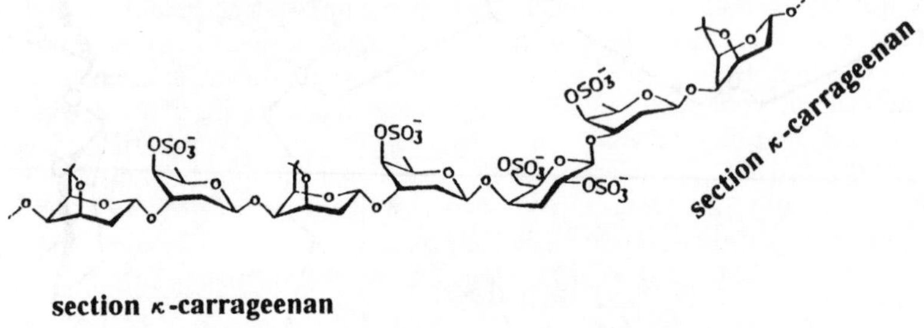

section κ-carrageenan

Figure 4   Kink between two κ-carrageenan segments. Only the smooth sections are capable of side-by-side association. Crystallization is prevented by the kinks

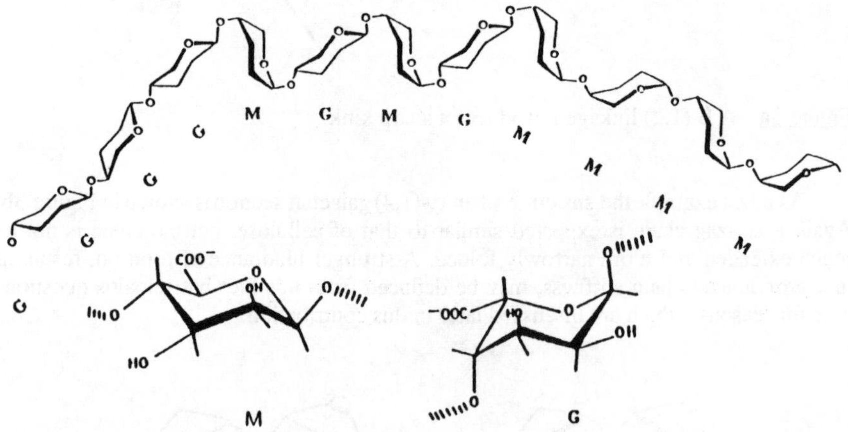

Figure 5   A section from an alginate chain. Only the polyguluronic acid sections are capable of lateral association but not the alternating $(MG)_x$- and $(MM)_y$- sections. The chemical structure of the guluronic and mannuronic acid units is shown as insert

The irregularity in the κ-carrageenan is caused by a unit in which the (3,6)-anhydro ring is broken and the corresponding C6-and C3 position is sulphated[4] (see Figure 4). The ring conformation now becomes unstable and the sugar chair flips into the other more stable conformation in which the substituents are placed equatorially. This flip causes a kink in the chain. Similar behaviour is observed with the alginates (Figure 5). A break in the macromolecular suprastructure is introduced by the inserted alternating mannuronic (*M*)-guluronic (*G*) section. The effect in the $[\beta\text{-}(1,4)]_x$ - $[\beta(1,3)]_y$ polysaccharide is not so much that a kink is introduced by the change from $\beta\text{-}(1,4)$ to $\beta\text{-}(1,3)$ blocks, but the two sections have two different suprastructures of non-uniform length, and this prevents macroscopic crystallization.

Cellulosic chains tend to crystallize and become water insoluble resulting in a strong self supporting structure which is needed in the primary cell wall. The introduction of any irregularity will strongly prevent such crystallization, and evidently Nature makes use of this fact to build up suprastructures which should not form the rigid structures of crystals. The tendency for aggregation or association is not fully suppressed as long as the regular sections are long enough. Clearly the smooth section in carrageenan or agar-agar can easily form bundles (compare Figure 6) in which the chains are side-by-side or laterally associated or aggregated. They form the junction zones in the thermally reversible gels. Such gels are highly swollen and display rubber elasticity which arises from the chain sections which are not incorporated in the junction zones. A similar effect is present in the alginates. Only the guluronic sections are capable of lateral association via Ca^{++} ions but not the alternating $(MG)_x$- and the $(MM)_y$- sections[4,5].

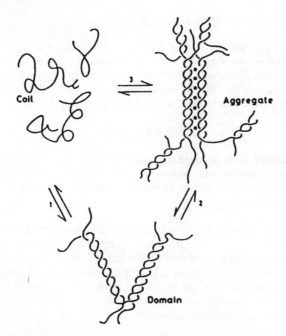

Coil ⇌ Aggregate

Domain

Figure 6 Model for side-by-side association for carrageenans and alginates[4,6]. A double helix formation is thought to preceed the association

There is presently much discussion as to how many chains are involved in the junction zones. The bundles have a considerable thickness in agar-agar as could be shown by electron microscopy and X-ray scattering[7]; for the other examples only a few chains are assumed to form the junctions, where double helices formation is considered as the essential starting condition for lateral aggregation. This brings up the question how such structures can be determined.

4 DETERMINATION OF MACROMOLECULAR   DIMENSIONS BY LIGHT   SCATTERING

In principle macromolecular architectures can be determined by the following three methods:

- common, or static light scattering,
- dynamic light scattering, and
- measurement of the intrinsic viscosity.

In former days the intrinsic viscosity data have been mostly used for characterization of macromolecular or colloidal structures in solution. This quantity is easily obtained from simple capillary viscosity measurements and was first introduced by Staudinger. It is given by the following equation

$$[\eta] = \{(\eta - \eta_o)/(\eta_o c)\} \text{ at } c = 0 \tag{1}$$

where this limiting value is obtained from the extrapolation of the term in curled brackets to zero concentration. The intrinsic viscosity is related to the molar mass of the particles and their dimensions, which may be characterized by a radius R_g to be defined later in this section. The following relationship goes back to Kirkwood and Riseman[8] and to a simplification by Fox and Flory[9] and Flory[10].

$$[\eta] = \phi (R_g^3/M) \tag{2}$$

This relationship gives good results for linear chains for which the factor ϕ is well defined. The prefactor depends, however, on hydrodynamic interaction that is influenced by the molecular architecture. In order to determine ϕ, measurements of M and R_g have to be performed.

Such measurements can be carried out by static light scattering (LS) which allows among others the determination of the weight average molar particle mass M_w and the radius of gyration R_g which is the root mean square distance of the various scattering elements (monomeric units, or chain segment) from the centre of mass[11] (Figure 7a). This geometrically defined average radius has to be distinguished from the hydrodynamic radius R_h which is obtained from the translational diffusion coefficient D using the Stokes-Einstein relationship as definition.

$$R_h = k_B T/(6\pi \eta_o D) \tag{3}$$

The translational diffusion coefficient can be measured by dynamic LS[12]. Figure 7b demonstrates schematically how the hydrodynamic radius is defined. If an open, only loosely coiled structure is present the solvent can penetrate fairly deeply into the coil and will drain through the outer section of the coil. There will be left only a small volume in the centre of the coil that will not be drained when the particle is moving. On the other hand, if the segment density in the coil is high, due to branching or a coil collapse, there will be only little draining at the periphery of the particle and the hydrodynamically effective radius will be much larger than in the first case of stiff chains.

Molecular structures in solution can be determined by two routes:

(a) the molar mass dependence of the two mentioned radii can be studied and eva-luated. This method requires a series of particles of the same architecture but dif-ferent molar masses.

(b) the ratio of the two radii $R_g/R_h = \rho$ can be measured for the same particle. This ratio is high ($\rho > 2.0$) for semiflexible chain behaviour and low for spheres ($\rho = 0.778$) or other globular structures. The random coil of linear chains lies in be-tween these values ($\rho = 1.5$-1.78, depending on the solvent quality), and its value decreases if branching is present[13].

Both techniques will be applied in the following.

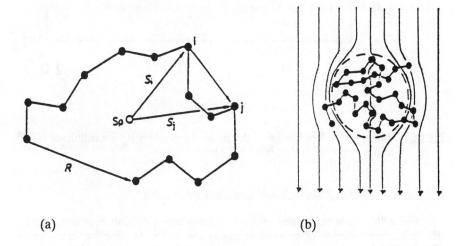

(a)   (b)

Figure 7   (a) Two dimensional representation of a random coil. The radius of gyration is defined as $R_g = [(1/N)\Sigma < S_j^2 >]^{1/2}$. (b) Representation of the hydrodyna-mic effective radius. The solvent only penetrates through the outer sections of the coil. A central part remains undrained; it defines the hydrodynamic radius

5 RESULTS

Cellulose and Amylose-Tri-Carbanilate

By the reaction of these polysaccharides with phenylisocyanate in hot pyridine the phenylurethane derivatives are obtained where all OH-groups are substituted. These tri-carbanilates are readily soluble in many organic solvents. Since all OH-groups are substituted there will be no H-bond formation which is responsible for the character-istic intramolecular suprastructure and for the probable intermolecular aggregation or association. Thus the sole influence of the different glucosidic bond is expected.

Figure 8 shows the result of static LS measurements for the two isomeric chains cellulose and amylose in acetone[14]. As expected the β-(1,4) linkage causes a more expanded conformation than the α-(1,4) one. A detailed examination reveals a slightly higher hindrance of rotation for the amylosic than for the cellulosic chain.

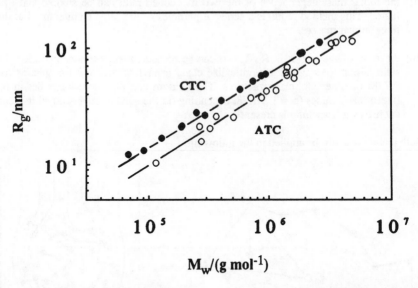

$$M_w/(g \ mol^{-1})$$

Figure 8   Molar mass dependence of the radius of gyration R_g for cellulose- (CTC) and amylose-tri-carbanilate (ATC) in acetone

Solution Behaviour of Partially Substituted Cellulosic Chains

The study of non substituted cellulose and amylose is difficult as strong aggregation (e.g. retrogradation of amylose in water) can occur. Much effort has been invested to disclose the structures formed on aggregation, so far with only partial success[15-17].

There are known, however, a large number of chemical derivatives of cellulose and one natural polymer is found in the seed of the tree Tamarind Indica. Figure 9 shows the chemical structure which has not to be taken as a repeating unit similar to those in bacterial polysaccharides but represents only the average composition of xyloses and galactoses attached to the backbone. The tamarind seeds polysaccharide (TSP) is soluble in water. The various cellulose derivatives studied so far are soluble in water or in organic solvents and sometimes in both. A fairly comprehensive report has recently been given and a complete documentation is not reproduced here[18].

Behaviour of common linear chains was expected, in particular

(1)   the hydrodynamic radius R_h should be proportional to the radius of gyration R_g.
(2)   both radii should exhibit power law behaviour, $R_g \sim R_h \sim M_w^\alpha$ with an exponent of $y = 0.5$ to 0.6, depending on the solvent quality.
(3)   the ratio $\rho = R_g/R_h$ should have values around 1.5 and should be independent of M_w[13].

None of these three expectations were found.

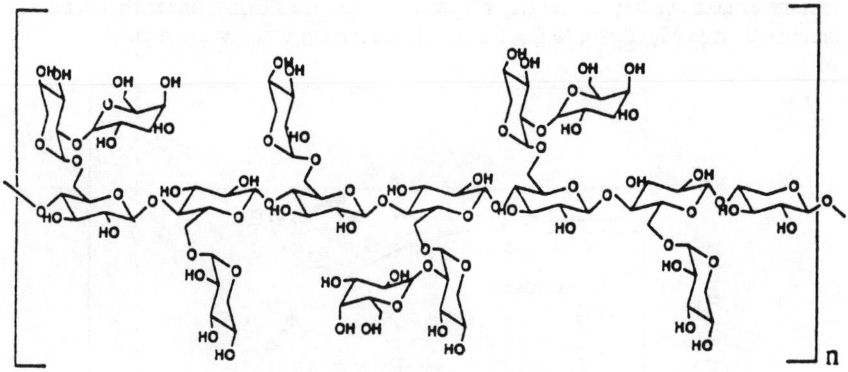

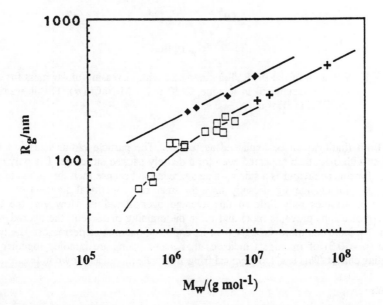

<u>Figure 9</u> Average chemical structure of TSP. The cellulose backbone is partially sub-
stituted by xyloses and partially by (1,2) galactosyl xyloses. The xylose substituents are bound by (1,6) links

<u>Figure 10</u> Molar mass dependences of the radii of gyration for carboxymethyl cellu-
lose (CMC) (◆), Na-alginate (+) and TSP (□) in 0.1 M NaCl aqueous solution and water respectively

Figure 10 exhibits the M_w dependence of R_g for a few examples and Figure 11
gives the M_w dependence of the ρ-parameter. Surprisingly the data from the various
derivatives show very similar behaviour. (i) At low M_w the radius of gyration increases
as expected but in the limit of large M_w the exponent decreases down to
α = 0.2[19], which is much smaller than α = 0.33 predicted for hard spheres. (ii) The ρ-

parameter takes values at low M_w which are typical for flexible linear chains but it decreases at large M_w down to $\rho = 0.3$ which is characteristic for microgels[20].

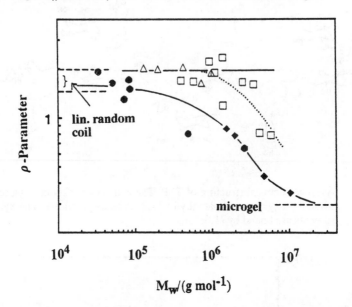

<u>Figure 11</u> Variation of the $\rho(=R_g/R_h)$ parameter with increasing molar mass for cellulose 2.5-acetate (o) in acetone, CMC in 0.1 M NaCl (♦), TSP in water (□) and CTC in Dioxan (Δ)

Both findings can be explained as follows. The particle radius increases much weaker with mass than expected even for a densely packed structure. The only explanation for this behaviour is a side-by-side aggregation because then the radius of gyration does not change significantly since the small cross-sectional diameter of such a bundle contributes only little to this average over-all radius. However, the dense packing of chains prevents more and more the draining of solvent, the hydrodynamic radius R_h increases more strongly than R_g and the ρ-parameter decreases. The typical value ($\rho = 0.3$) of microgels indicates that many chains are bundled together with dangling chains. This leads to a typical fringed micelle model as shown in Figure 12.

Xylans

Xylans from various sources have been studied[21]. They are designated as GX, AGX and AX, where X stands for the xylan backbone, G for the (1,2) linked glucuronic acid, and A for arabinofuranose which can be attached to the backbone in C2 or C3 position or both. All xylans appear to be soluble in dimethyl sulphoxide (DMSO)/water mixtures of 90/10 (v/v). Figure 13 shows the results of static light scattering[22]. Extremely high molar masses (up to $50 \cdot 10^6$) and no power law behaviour were observed. Evidently a strong aggregation is present. The structure of these aggregates could not be determined since, so far, only static LS has been carried out. Discrimination between various models will be possible when dynamic LS is performed and the hydrodynamic radius is measured.

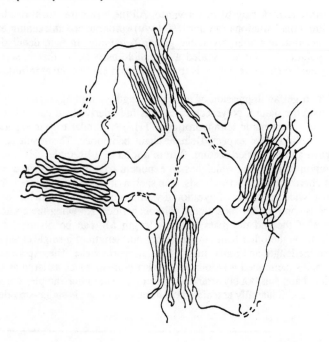

Figure 12 Fringed micelle model that describes the findings of Figure 10 and 11

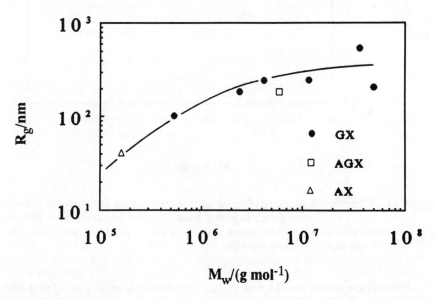

Figure 13 Molar mass dependence of R_g for xylans from various sources in DMSO/H_2O (90/10)

One further remark may be of relevance. All the structures mentioned display at even fairly low concentrations (around 1% (w/v)) pronounced thickening effects and at higher concentrations often reversible gelling behaviour. In fact, detailed examination of the solution structure revealed a reversible association that is superimposed onto the aggregated structure, which is irreversible and exists even at infinite dilution.

In order to get an impression of the chain length of non-aggregated xylan chains we transformed the natural polysaccharide in its fully derivatized carbanilate and performed LS measurements in tetrahydrofuran (THF)[23]. Similar to our experience with amylose and cellulose we expected single chain behaviour. The result as shown in Figure 14 gave an unexpected picture: 9 from a total of 14 samples fall approximately on one common straight line which has an exponent of a little less than $\alpha = 1.0$. This exponent is characteristic for rigid rods. For such a structure the contour length L of the rod is just twice the radius of gyration if we presume a most probable length distribution which is common for most linear polysaccharides. Knowing the contour length, and the mass of the rod the mass per unit length M_L can be obtained from these measurements. On the other hand, the mass per unit length of a single chain can be estimated from the length and molar mass of the anhydroxylose. The experimental value for the 9 samples mentioned was found to be very close to twice as large as calculated. Evidently the chains has doubly stranded structures. The other samples correspond to bundles of 3, 6 and 7 laterally associated chains and only one is single-stranded.

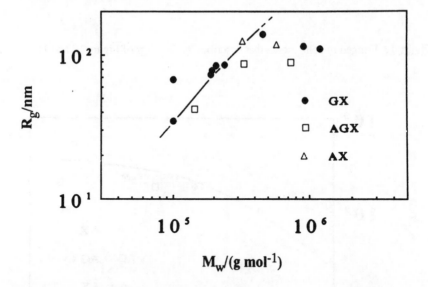

Figure 14 Molar mass dependence of R_g for carbanilated xylans. The drawn line corresponds to a wormlike chain with a Kuhn segment length of about 650 nm which is about 5 times larger than the contour length of the chain. The chains behave virtually like rigid rods

Probably the double strand formation has a rather simple explanation. The uronic acid groups in GX and AGX are not substituted by phenylisocyanate. In organic solvents the free acid groups can form strong H-bonds with another chain which will stiffen the chain remarkably. The single stranded chain may result from a glucurono xylan in which all glucuronic acid groups have been cleaved.

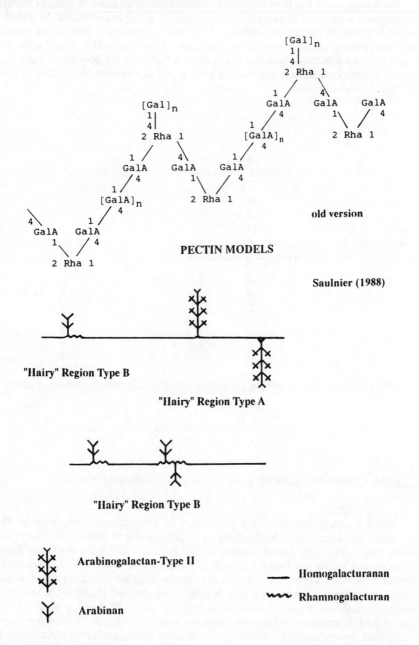

old version

PECTIN MODELS

Saulnier (1988)

"Hairy" Region Type B

"Hairy" Region Type A

"Hairy" Region Type B

Arabinogalactan-Type II

Arabinan

Homogalacturanan

Rhamnogalacturan

Figure 15 Pectin models

Pecularities of Pectin

For a long time it has been believed that pectin consists mainly of partially methyl-ated α-galacturonic acid blocks which after about 15 units are interrupted by one or a few (1,2) linked rhamnose molecules[24]. This sugar introduces a kink in the regular chain and prevents crystallization of the polygalacturonic chain. This perturbation may be amplified by a further chain that is linked as a branch in C4 position of the rham-nose. This model would fit nicely into the general picture of reversible gelation of plant polysaccharides[5,6].

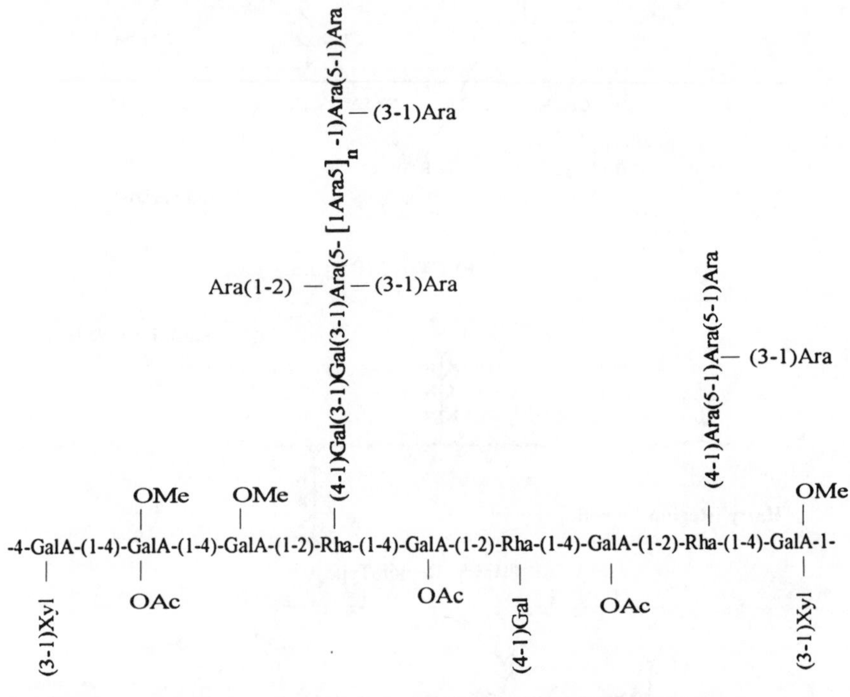

Figure 16 Chemical structure of "hairy" regions as postulated by Voragen[26]

Recent investigations disclosed, however, a far more complicated structure. This conclusion has been drawn mainly from two essential observations. The first is that besides rhamnose, other neutral sugars, mainly galactose and arabinose were found. These neutral sugars are linked together to form a highly branched macromolecular domain. Figure 15 shows the old and new pectin models[25]. At least two types of "hairy" regions are present. One type has been investigated in detail by Voragen[26] (Figure 16) and was recently confirmed by Will[27] (Figure 17). Another structure con-sists of β-(1,3) galactan- and β-(1,6) galactan- sections which at the periphery carry (1,3) linked arabinofuranose units. The two kinds of chains are linked together via β-(1,6) or β-(1,3) bonds respectively to form a highly branched structure.

We carried out LS measurements from the same fraction[29] of an arabinogalactan that was isolated from apple juice and analysed by Will with GPC. Instead of M_w =

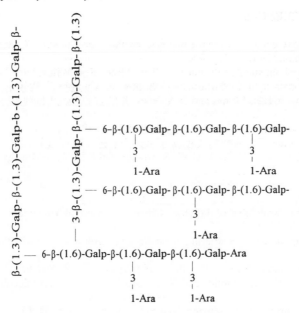

Figure 17 Chemical structure of "hairy" regions as postulated by Will et al.[27]

50000, direct light scattering gave M_w = 200000, and a ρ-parameter which is characteristic of globular or densely branched material[28]. Thus, the hairy regions are not small, and the large neutral sugar domains very likely have special functions in the complex molecule. The prevailing opinion, that only the polygalacturonic section induces the desired gelation, has probably to be revised. Recent measurements by ourselves give strong evidence for an active part of these domains in the association and gelation behaviour[29].

6 CONCLUSIONS

The last example of pectin demonstrates how complex the spatial structures can be. Similar problems are expected with amylopectin and its interaction with amylose. In spite of these complexities it becomes clear that the universality of laws as found for non-polar synthetic polymers are not completely lost by the numerous possibilities of association, but the type of association or aggregation is predetermined by the conformation of the individual chains.

Regular chemical composition would result in crystalline fibrous structures of little flexibility. Crystallinity and the resulting rigidity is prevented by introduction of perturbations either directly into the main chain or by regio specific derivatization. In the latter case unsubstituted sections are obtained beside others with a higher degree of substitution than the average. Only the smooth sections are capable of association which eventually leads to gelation or to a partially ordered fringed micelle. The "hairy" regions in pectin present large structural heterogeneities and have additional not yet fully understood functions in native tissues.

7 REFERENCES

1. P.-G. De Gennes, 'Scaling Concepts in Polymer Physics', Cornell University Press, Ithaca, New York, 1979.
2. P.J. Flory, 'Statistical Mechanics of Chain Molecules', Wiley, New York, 1969.
3. J.R. Woodward, D.R. Phillips and G.B. Fincher, Carbohydr. Polym., 1988, 8, 85.
4. G. Robinson, E.R. Morris and D.A. Rees, J. Chem. Soc., Chem. Commun., 1980, 152.
5. O. Smidsroed and H. Grasdalen, Carbohyd. Res., 1982, 2, 270.
6. D.A. Rees and E.J. Welsh, Angew. Chem. Int. Ed., 1977, 16, 214.
7. M. Djarbourov, A.H.Clark, D.W. Rowlands and S.B. Ross-Murphy, Macromolecules, 1989, 22, 180.
8. J.G. Kirkwood and J. Riseman, J. Chem. Phys., 1948, 16, 565.
9. T.G. Fox and P.J. Flory, J. Am. Chem. Soc., 1951, 73, 1909.
10. P.J. Flory, 'Principles of Polymer Chemistry', Cornell University Press, Ithaca, New York, 1953.
11. M.L. Huglin, 'Light Scattering from Polymer Solutions', Academic Press, London, 1972.
12. B.J. Berne and R. Pecora, 'Dynamic Light Scattering', Wiley, New York, 1976.
13. W. Burchard, M. Schmidt and W.H. Stockmayer, Macromolecules, 1980, 13, 580.
14. W. Burchard and E. Husemann, Makromol. Chem., 1961, 44-46, 358.
15. E. Husemann, B. Pfannemüller and W. Burchard, Makromol. Chem., 1963, 59, 16.
16. W. Burchard, Makromol. Chem., 1963, 59, 110.
17. M.J. Gidley, P.V. Bulpin, A.H. Clark, R.K. Richardson and S.B. Ross-Murphy, Macromolecules, 1989, 22, 341; 345; 351.
18. W. Burchard and L. Schulz, Papier, 1989, 43, 663.
19. P. Lang, Ph.D. Thesis, Freiburg, 1990.
20. M. Schmidt, D. Nerger and W. Burchard, Polymer, 1979, 20, 582.
21. A. Ebringerova, Z. Hromadkova, W. Burchard, W. Vorwerg and G. Berth, Abstract 8th Int. Bratislava Symp., 1991.
22. A. Ebringerova., W. Vorwerg and W. Burchard, Manuscript in preparation.
23. P. Merz, Diploma Thesis, Freiburg, 1990.
24. G.O. Aspinall, 'The Polysaccharides', Academic Press, London, 1982, Vol 1, p.263ff.
25. L. Saulnier, J.M. Brillouet and J.P. Joseleau, Carbohyd. Res., 1988, 181, 1.
26. F.G.J. Voragen and W. Pilnik, J.Am. Chem. Soc., 1989, 7, 93.
27. F. Will and H. Dietrich, Carbohyd. Polym., 1992, 18, 109.
28. F. Will, H. Dietrich, B. Hässlin and W. Burchard, Manuscript in preparation.
29. B. Hässlin, Ph.D. Thesis, Freiburg, 1992.

STRUCTURE AND FUNCTION OF BARLEY MALT α-AMYLASE

M. Søgaard[1], A. Kadziola[2], J. Abe[1], R. Haser[2] and B. Svensson[1]

[1] DEPARTMENT OF CHEMISTRY, CARLSBERG LABORATORY,
2500 COPENHAGEN, DENMARK
[2] LLCMB-CNRS, URA 1296, 13326 MARSEILLE, FRANCE

1   INTRODUCTION

α-Amylases catalyze the hydrolysis of internal α-(1→4)-glucosidic linkages in starch and related poly- and oligosaccharides. Structural details on α-amylase-substrate interactions emerged with the X-ray crystallographic structure analysis of maltose and a maltotriose analogue in complex with Taka-amylase A[1] and porcine pancreatic α-amylase[2], respectively. In kinetic experiments the different α-amylases have been found to possess extended substrate binding areas that span from five to ten binding subsites each accommodating a substrate glucosyl residue[3-6].

Architecture

α-Amylases are Ca^{2+}-containing[1,2,7-9] multidomain proteins that belong to the $(ß/α)_8$-barrel protein family[1,2,9]. As is the case for all enzymes having this fold pattern, the substrate binding site is created by residues at the C-terminal end of ß-strands, forming the cylindrical inner ß-sheet, and the helix-connecting loops extending from such strands[10]. In α-amylases and structurally related amylolytic enzymes a long third loop of the barrel folds in a small separate structural domain[1,2,9,11] that, stabilized by the structural Ca^{2+}, participates in substrate binding[1,2]. Variation in amino acid sequence and loop length in the regions involved in substrate binding is very important in determining the number of binding subsites and the substrate specificity[12-15]. Structure prediction procedures applied to amino acid sequences of enzymes in the α-amylase family have enabled sequence alignment guided by secondary structure elements as a way to study substrate binding loop regions and correlate specific structural features with the different action patterns and substrate specificities[12-15]. Finally, a picture of the different domains and their organization along the polypeptide chain of amylolytic enzymes tentatively suggests specific motifs associated with certain substrate specificities[16]. Different debranching and branching enzymes, responsible for hydrolysis and synthesis, respectively, of α-(1→6)-linkages in amylopectin and glycogen, thus show characteristic features both at the domain level and in putative binding areas of the catalytic domain[15,16].

Mechanism

α-Amylases presumably work by a double displacement, or S_N2, mechanism that involves the nucleophilic attack of an enzyme carboxylate at C-1 of the terminal residue of the substrate glycon part to yield a covalent ß-glucosyl enzyme intermediate[17-19]. As in all mechanisms proposed for carbohydrases, a general acid catalyst is crucial for initial protonation of the glucosidic oxygen of the substrate bond to be cleaved. Here its base form later activates, in the reverse reaction, a water molecule for nucleophilic attack at C-1 of the enzyme-substrate covalent intermediate. Experimental evidence for this catalytic pathway has been provided by demonstration of a covalent intermediate between maltotetraose and porcine pancreatic α-amylase by the aid of nuclear magnetic resonance spectroscopy at low temperature[19].

Catalytic Site

Identification of amino acid residues involved in substrate binding and catalysis is best done by a combination of crystallographic, chemical, genetic, and enzyme kinetic techniques. A recent review that addresses the mutational analysis of glycosylase function discusses the currently available data[20].Crystallographic studies on α-amylases from *Aspergillus oryzae* (Taka-amylase A)[1], porcine pancreas[2], *Aspergillus niger*[9], and, most recently, barley[21] have pointed out three invariantly occurring carboxylic acid residues Asp 206, Glu230, and Asp297 (Taka-amylase A numbering) at or near the site of catalysis. One of these is the general acid/base catalyst and one is the enzyme nucleophile, that is suggested to undergo glucosylation by the glycon part of the substrate during catalysis. Three nearby histidyl residues, the invariant His122 and His296 and the less well-conserved His210 (Taka-amylase A numbering), are likely to play important roles in the mechanism[1,2].

Starch Binding Site

Many amylolytic enzymes, including α-amylases, have affinity for starch granules. In some cases a starch binding site is associated with a separate structural domain different from the catalytic domain[11,22-24], while in other ones a secondary binding site has been found on the catalytic domain at a distance from the active site[25,26]. Generally protein-carbohydrate interactions involve aromatic residues[27,28], tryptophans are thus found at such secondary binding sites in α-amylase from barley[26], glucoamylase from *Aspergillus niger*[29,30], and a cyclodextrin glucanosyltransferase from *Bacillus stearothermophilus*[31].

Barley α-Amylase

In barley, two gene families encode the different forms of low pI α-amylase (AMY1) and high pI α-amylase (AMY2)[32]. Between the families approx. 80% amino acid residue identity is found[33,34], while 95% or higher identity exists for members of the same family[32]. In spite of this high structural similarity, AMY1 and AMY2 display distinct differences in Ca^{2+}-affinity[8], sensitivity to sulfhydryl and chelating reagents[7], stability at acidic pH values and elevated temperatures[7], activity on starch granules[35], affinity for soluble starch and oligodextrin substrates[36-38], and sensitivity to the endogenous α-amylase/subtilisin inhibitor (BASI)[39,40]. These isozymes therefore represent a very attractive system for correlation of specific structural features with specific physico-chemical, chemical, and enzymic properties. Crystals suitable for X-ray diffraction analysis have been prepared of AMY2[41], the dominating isozyme in malt. Very recently a model of the three-dimensional structure has been constructed for a member of that isozyme family, AMY2-2[21].

We have established a heterologous expression system for two cDNAs encoding an AMY1 and an AMY2 family member, respectively[37]. By utilizing this system, recombinant single mutants have been prepared of AMY1 to investigate the roles in the mechanism of action of His93, Asp180, Glu205, His290, and Asp291[42], the five invariant active site residues mentioned above. Mutation is in addition performed at Trp278 and Trp279[42], two residues previously identified by chemical modification to interact with cycloheptaamylose[26] at a secondary binding site presumably overlapping with the site for binding onto starch granules[43].

2   RESULTS

Three-Dimensional Structure of Barley High pI α-Amylase

Crystallization.   The highly purified form of barley high pI α-amylase, AMY2-2 (pI 5.93), was prepared according to the usual protocol from kilned malt involving fractionated ammonium sulfate precipitation of the flour extract, affinity chromatography on cycloheptaamylose-Sepharose, and CM-Fractogel ion exchange chromatography[44] followed by separation of individual components of the AMY2 family by preparative chromatofocusing[38]. Crystallization was done by the hanging drop technique essentially as reported for AMY2 at pH 6.7 in 0.05 M MES, 5 mM $CaCl_2$ using $(NH_4)_2SO_4$ as precipitating agent[41]. Another major form, AMY2-1 (pI 5.88)[38] was much less soluble than AMY2-2 and did not crystallize under the same conditions.

Crystallographic Structure Analysis.   The tertiary structure of AMY2-2 has now been solved on the basis of single crystal X-ray diffraction data collected with a multiwire area detector to 2.8 Å resolution for the native crystal and three heavy-atom derivatives[21]. The complete polypeptide chain of 403 residues could be fitted to the electron density map and the model refined using the molecular dynamics program X-PLOR.

This is the first three-dimensional model of an α-amylase from a higher plant. It includes 403 amino acid residues, 162 well-defined water molecules and three Ca^{2+} binding sites. Like the known structures of other α-amylases[1,2,9], it is organized in three domains: a $(ß/α)_8$-barrel domain of about 270 residues (domain A), with a small domain of about 70 residues protruding from the barrel between the third ß-strand and the third α-helix (domain B), and a C-terminal domain of about 60 residues folded in a characteristic five stranded antiparallel ß-sheet (domain C). About 75% of the crystal volume is solvent distributed mainly in huge channels running through the lattice. Diffusion of substrate analogues into the native crystals has recently allowed identification of specific binding sites for oligosaccharides interacting with the enzyme.

Ca^{2+}-Binding.   The three Ca^{2+} ions are liganded to domain B in an unusual manner. Two Ca^{2+} thus interact with the same aspartic acid residue. The third Ca^{2+} is in a site not hitherto reported for enzymes from the α-amylase family and seems to be more loosely bound, since it can be substituted selectively by Eu^{3+}[21].

AMY2 is much more sensitive to Ca^{2+} than AMY1[7,8]. We have therefore reinvestigated the influence of the concentration of Ca^{2+} on the catalytic activity of the different forms of barley α-amylase, AMY2-1, AMY2-2, and AMY1. Characteristically the hydrolytic activity towards blue starch of both of the AMY2 forms increased approx. three fold with the $CaCl_2$ increasing from 1-15 mM. The activity of AMY2 further decreased when the $CaCl_2$ was reduced in the range from 1 to 0.1 mM. The AMY1 activity throughout the entire range remained essentially constant at a level corresponding to the value obtained with the AMY2 forms at about 0.5 mM $CaCl_2$. The activity of both AMY1 and AMY2, however, dropped dramatically from 15 to 40 mM $CaCl_2$[45]. We assume this effect at high concentration of $CaCl_2$ is due to Ca^{2+} interacting with carboxylic acid groups at the active site, resembling the type of Ca^{2+} binding reported for the acid and neutral α-amylases from *Aspergillus niger*[9]. Since the crystals of AMY2-2 have been grown at 5 mM of Ca^{2+}, we thus expect that a fourth and inactivating Ca^{2+} can bind to AMY2-2 in addition to the three Ca^{2+} depicted in the current model.

Mutational Studies of the Function of Barley α-Amylase

Recombinant Forms. cDNAs encoding AMY1 and AMY2 have been expressed in yeast using the vector pMA91 that contains the phosphoglycerate kinase promoter and terminator[37]. Yeast was chosen as host because active recombinant wheat and mouse salivary α-amylases had previously been produced in that organism[46,47]. Recombinant AMY1 and AMY2 are both secreted to the medium as directed by the barley signal peptides. The recombinant enzymes are easily purified by affinity chromatography on cycloheptaamylose-Sepharose[37]. SDS-PAGE and amino acid sequencing confirmed the size and the N-terminal sequence for AMY1, while AMY2 had a blocked N-terminus, presumably a pyroglutamate residue as found for the AMY2 isolated from malt[48]. Since the yield of recombinant AMY1 was 1-3 mg per liter, while only approx. 1/20th of that amount was produced of AMY2, it was decided to make site-specific mutants initially in AMY1 even though this isozyme has not yet been crystallized to give crystals suitable for X-ray diffraction studies[41] and AMY2 is the dominant isozyme in malt.

A slight decrease was observed in the specific activity of recombinant AMY1 and isoelectric focusing revealed that four dominant bands, two with wild-type level and two with low activity in zymograms, were produced from the single cDNA[37]. The cause for this unexpected processing had to be understood prior to the mutagenesis experiments in order to be able to produce a single type of polypeptide chain. Previous observations in barley aleurone cell cultures had revealed a precursor-product relationship for AMY1, two transient forms of higher pI were residing mainly intracellularly and two processed forms were found in the medium[49]. Since in that case phosphorylation, sulfation, acetylation or glycosylation were ruled out as possible explanations for the modification, we resorted to either proteolysis or deamidation as being the responsible factor. Testing this hypothesis by addition of carboxypeptidase Y to the recombinant AMY1 preparation readily indicated that the two forms of higher pI were converted into the two forms of lower pI[50]. However, the corresponding zymogram still revealed that one of these forms had very low activity. Careful amino acid analysis of the individual homogeneous forms purified from the culture medium then suggested that glutathione was present in those showing low activity. Reduction subsequently confirmed this idea as seen by an increase in the pI value, regain of the activity, and demonstration of glutathione release[50]. Recently electrospray mass spectrometry of the four individual forms strongly supported that the carboxypeptidase treatment liberates the C-terminal Arg-Ser from the two forms of unusually high pI and that the two forms of low activity contain one molecule of glutathione[51]. The carboxypeptidase trimming of the malt AMY1 actually resulted in liberation of about seven amino acid residues from the C-terminal part. This processing was observed in vivo studies on aleurone cell protoplasts to be suppressed by addition of exogenous serine carboxypeptidase inhibitors[50]. The incomplete post-translational modification of recombinant AMY1 in yeast was explained by the inefficient action of the Kex1 protease, a Golgi-localized serine carboxypeptidase of carboxypeptidase B-like specificity, as well as of other carboxypeptidases specific for the residue types relevant in processing of AMY1[50]. Recombinant AMY1 moreover lacks a disulfide bridge involving Cys95 in malt AMY1. Since this residue is substituted in AMY2, we speculate that Cys95 is the site of glutathionylation adversely affecting the activity of recombinant AMY1. Cys95 was therefore mutated to Ala. As a consequence a homogeneous and fully active polypeptide was produced when in addition the codons for the C-terminal Arg-Ser had been removed[51]. Thus a genetic solution is outlined to the problem of multiple recombinant forms of AMY1 of varying activity. Ideally this is the cDNA construction to use in mutational investigation of structure/function relationships in AMY1. The mutant AMY1 forms described below, however, were either treated with DTT to eliminate the glutathione or they have been compared with a reference of recombinant wild-type AMY1 without DTT pretreatment.

Active Site Mutants. The significance in activity of five invariant amino acid side chains, that either interact with the terminal glycon-part glucosyl unit of the substrate bond to be cleaved or are situated in the near vicinity, has been investigated by applying a site-directed mutation approach to the AMY1 cDNA followed by expression in yeast. The yields greatly varied, thus wild-type level was found for Asp291 →Asn, while 10% of that was obtained for Asp180→Asn and Glu205→Gln. The two histidine mutants, His93→Asn and His290→Asn, were present in about 1% of the wild-type recombinant AMY1. When tested on oligo- and polysaccharide substrates, the three carboxylic acid mutants were devoid of activity, while His93→Asn retained 5% and His290→Asn 10% of wild-type activity towards insoluble blue starch[42]. The smaller the substrate, the greater the loss of activity, suggesting that dissociation of the enzyme-substrate complex competed significantly with hydrolysis for these AMY1 mutants. Since K_m was affected only to a small extent the two histidyl residues are believed to be involved in transition state stabilization. Further support for this interpretation is provided by the fact that the binding affinity (K_d) for acarbose, an inhibitor and transition state analogue, is weakened to the same extent as the specificity constant (k_{cat}/K_m) is reduced[42]. Also the pK_2 of both of the free mutant enzymes was essentially unchanged confirming that the histidyl residues are not directly involved in the substrate conversion as general acid/base catalyst[42].

Secondary Site Mutation. The Trp278 and Trp279 were previously identified to be involved in cycloheptaamylose binding outside of the active site[26]. Expression of mutant cDNAs encoding the double mutant and the two single mutants where Trp was replaced by Ala resulted in detectable protein only for Trp279→Ala AMY1, which was produced in about 1% of wild-type. It retained about 22% of the wild-type activity, while the affinity for granular starch, as given by K_d, decreased from 0.2 mg/ml for wild-type to 2 mg/ml for the mutant[42].

3 DISCUSSION

The present mutational studies on barley α-amylase 1 (AMY1) have demonstrated that the three invariant carboxylic acid residues are essential for activity and that the two invariant histidyl residues are important but not directly involved in catalysis. Furthermore, two adjacent tryptophanyl residues located at a distance from the active site are confirmed to play a significant role in the adsorption onto starch granules.

These findings are fully supported by recent crystallographic evidence of the interaction between acarbose and barley α-amylase 2 (AMY2) shown to occur both at the active site as well as at the secondary site[52]. However, the present mutations do not allow assignment of the catalytic general acid/base and enzyme nucleophile to specific carboxylic acid residues. We speculate that Glu205, which is located in a more hydrophobic environment[1,2], is the proton donor and that a weak sequence similarity around Asp180 in AMY1 and mechanism-based affinity-labelled acid residues from distantly related enzymes[53-56] makes Asp180 a candidate for the enzyme nucleophile.

This work has been supported by grants from the E.E.C. Biotechnology Action Programme to R.H. and B.S. and by a pre-doctoral fellowship from the Carlsberg Foundation to A.K.

4   REFERENCES

1. Y. Matsuura, M. Kusunoki, W. Harada and M. Kakudo, J. Biochem. (Tokyo), 1984, 95, 697.
2. G. Buisson, E. Duée, R. Haser and F. Payan, EMBO J., 1987, 6, 3909.
3. J.F. Robyt and D. French, Arch. Biochem.Biophys., 1963, 100, 287.
4. C. Seigner, E. Prodanov and G. Marchis-Mouren, Eur. J. Biochem., 1985, 148, 161.
5. J.F. Robyt, Denpun Kagaku, 1989, 36, 287.
6. E.A. MacGregor and A.W. MacGregor, 'New Approaches to Research on Cereal Carbohydrates', Elsevier, Amsterdam, 1985, p. 149.
7. E. Bertoft, C. Andtfolk and S.-E. Kulp, J. Inst. Brew., 1984, 90, 298.
8. D.S. Bush, L. Sticher, R.B. van Huystee, D. Wagner and R.L. Jones, J. Biol. Chem., 1989, 264, 19392.
9. E. Boel, A.M. Brzozowski, Z. Derewenda, G.G. Dodson, V.J. Jensen, S.B. Petersen, H. Swift, L. Thim and H. Woldike, Biochemistry, 1990, 29, 6244.
10. G.K. Farber and G.A. Petsko, Trends Biochem. Sci., 1990, 15, 228.
11. C. Klein and G.E. Schulz, J. Mol. Biol., 1991, 217, 737.
12. E.A. MacGregor, J. Prot. Chem., 1988, 7, 399.
13. E. Raimbaud, A. Buléon, S. Perez and B. Henrissat, Int.J. Biol. Macromol., 1989, 11, 217.
14. E.A. MacGregor and B. Svensson, Biochem. J., 1989, 259, 145.
15. H.M. Jespersen, E.A. MacGregor, B. Henrissat, M.R. Sierks and B. Svensson, submitted.
16. H.M. Jespersen, E.A. MacGregor, M.R. Sierks and B. Svensson, Biochem. J., 1991, 280, 51.
17. D.E. Koshland, Biol. Rev., 1953, 28, 416.
18. E.H. Fischer and E.A. Stein, 'The Enzymes', Academic Press, N.Y., 1960, p. 389.
19. B.Y. Tao, P.J. Reilly and J.F. Robyt, Biochim. Biophys. Acta, 1989, 995, 214.
20. B. Svensson and M. Søgaard, J. Biotechnol., 1992, in press.
21. A. Kadziola, J. Abe, B. Svensson and R. Haser, in preparation.
22. B. Svensson, T.G. Pedersen, I. Svendsen, T. Sakai and M. Ottesen, Carlsberg Res. Commun., 1982, 47, 55.
23. B. Svensson, H.M. Jespersen, M.R. Sierks and E.A. MacGregor, Biochem. J., 1989, 264, 309.
24. N.J. Belshaw and G. Williamson, FEBS Lett., 1990, 269, 350.
25. F. Payan, R. Haser, M. Pierrot, M. Frey, J.P. Astier, B. Abadie, E. Duée and G. Buisson, Acta Cryst., 1980, B36, 416.

26. R.M. Gibson and B. Svensson, Carlsberg Res. Commun., 1987, 52, 373.
27. F.A. Quiocho, Ann. Rev. Biochem., 1986, 55, 287.
28. J.C. Spurlino, G.-Y. Lu and F.A. Quiocho, J. Biol. Chem., 1991, 266, 5202.
29. B. Svensson, A.J. Clarke and I. Svendsen, Carlsberg Res. Commun., 1986, 51, 61.
30. B. Svensson and M.R. Sierks, Carbohydr. Res., 1992, 227, 29.
31. M. Kubota, Y. Matsuura, S. Sakai and Y. Katsube, International Symposium on Cereal and other Plant Polysaccharides, Kagoshima, 1990, abstr. P 46.
32. R.L. Jones and J.V. Jacobsen, Int. Rev. Cytology, 1991, 125, 49.
33. J.C. Rogers and C. Milliman, J. Biol. Chem., 1983, 258, 8169.
34. J.C. Rogers, J. Biol. Chem., 1985, 260, 3731.
35. A.W. MacGregor and J.E. Morgan, Cereal Foods World, 1986, 688.
36. A.W. MacGregor, J.E. Morgan and E.A. MacGregor, Carbohydr. Res., 1992, 227, 301.
37. M. Søgaard and B. Svensson, Gene, 1990, 94, 173.
38. E.H. Ajandouz, J. Abe, B. Svensson and G. Marchis-Mouren, Biochim. Biophys. Acta, 1992, 1159, 193.
39. J. Mundy, J. Hejgaard and I. Svendsen, Carlsberg Res. Commun., 1983, 48, 81.
40. R.J. Weselake, A.W. MacGregor and R.D. Hill, Plant Physiol., 1983, 72, 809.
41. B. Svensson, R.M. Gibson, R. Haser and J.P. Astier, J. Biol. Chem., 1987, 262, 13682.
42. M. Søgaard, A. Kadziola, R. Haser and B. Svensson, in preparation.
43. R.J. Weselake and R.D. Hill, Cereal Chem., 1983, 60, 98.
44. R.M. Gibson and B. Svensson, Carlsberg Res. Commun., 1986, 51, 295.
45. M. Søgaard, J. Abe, M. F. Martin-Eauclaire and B. Svensson, 'Proceedings of Workshop on Cereal Polysaccharides', Le Croisic, 1992, in press.
46. K.K. Thomsen, Carlsberg Res. Commun., 1983, 48, 545.
47. S.J. Rothstein, K.N. Lahners, C.M. Lazarus, D.C. Baulcombe and A.A. Gatenby, Gene, 1987, 55, 353.
48. B. Svensson, J. Mundy, R.M. Gibson and I. Svendsen, Carlsberg Res. Commun., 1985, 50, 15.
49. J.V. Jacobsen, S.V. Bush, L. Sticher and R.L. Jones, Plant Physiol., 1988, 88, 1168.
50. M. Søgaard, F.L. Olsen and B. Svensson, Proc. Natl. Acad. Sci. U.S.A., 1991, 88, 8140.
51. M. Søgaard, J. Andersen, P. Roepstorff and B. Svensson, in preparation.
52. A. Kadziola, M. Søgaard, B. Svensson and R. Haser, in preparation.
53. B. Svensson, FEBS Lett.,1988, 230, 72.
54. W. Hunziker, M. Spiess, G. Semenza and H.F. Lodish, Cell, 1986, 46, 227.
55. G. Mooser, S.A. Hefta, R.J. Paxton, J.E. Shively and T.D. Lee, J. Biol. Chem., 1991, 266, 8916.
56. M.M.P. Hermans, M.A. Kroos, J. van Beeumen, B.A. Oostra and A.J.J. Reuser, J. Biol. Chem., 1991, 266, 13507.

USE OF EXTRUSION PROCESSES FOR ENZYMIC AND CHEMICAL MODIFICATIONS OF STARCH

G. Della Valle, P. Colonna, J. Tayeb and B. Vergnes[1]

INRA, CENTRE DE RECHERCHES AGRO-ALIMENTAIRES, BP527,
44026 NANTES CÉDEX3, FRANCE
[1] CEMEF, ECOLE NATIONALE SUPÉRIEURE DES MINES DE PARIS, BP 207
06904 SOPHIA ANTIPOLIS CÉDEX, FRANCE

1 INTRODUCTION

Apart from conventional food applications, there has been some efforts in the last ten years to consider the possible use of twin screw extruders as biochemical and chemical reactors. Compared to classical batch processes, extrusion offers the following advantages: it is a continuous, short time process and it can be run with low hydration conditions. Conversely, the lack of knowledge on the effect of high temperature and shear on reagents has slowed the development of such applications of extrusion technology. The aim of this work is to explain the limitations of extrusion process for enzymic conversion in terms of enzyme sensitivity to mechanical processing and to show how an extruder can be used as a multiple stage chemical reactor for starch modification.

2 ENZYMIC CONVERSION

The use of twin screw extrusion for starch saccharification by continuous enzymic conversion is not new and Linko[1] reported that the conversion needed to be performed in mild conditions in order to prevent enzyme deactivation. More recently, Roussel[2] achieved sequential gelatinization and liquefaction by injecting thermostable α-amylase after the first step. Results were obtained under the form of response surfaces to determine the optima DE (Dextrose Equivalent) as function of extrusion conditions (Figure1). For instance the influence of screw speed N and barrel temperature in the liquefaction section Tl was obtained for low feed rates (Q = 5 kg/h). Even if the DE is correct, the conditions of extrusion employed, especially the feed rate, make the process rather inefficient and this might be due to the unusual conditions of reaction of the enzyme submitted to high temperature and shear. Despite these practical results, there is a lack of knowledge on the applicability of enzymes in thermomechanical treatment.

The objective of this work was to obtain data on the magnitude and duration of shearing forces which enzyme molecules can withstand during starch gelatinization-melting in an extruder. For this purpose, we have used a pre-shearing rheometer, called Rhéoplast, to study separately the effects of each phenomenon in a controlled manner.

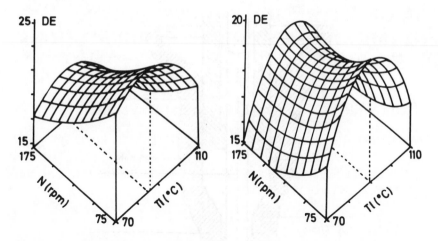

<u>Figure 1</u> Influence of screw speed and temperature on starch liquefaction at mois-
ture 20% (left) and 40% (right) from Roussel[2]

Material and Methods

<u>Starch Sample</u>. A commercial corn starch (Roquette Frères, Lestrem, France)
was processed in a Rhéoplast ® (Courbon SA BP327 F-42015 StEtienne) at two
moisture contents 40 and 50% (total matter basis), with water whose pH was adjusted to 6.5. Starch samples were water-equilibrated overnight before processing.
Alpha-amylase, Termamyl 60L, was a gift from Novo (Denmark).

<u>Processing</u>. The amylase catalyzed hydrolysis of starch was conducted in a
Rhéoplast viscometer. The initial concentration of enzyme in all experiments was
2 µl/g starch (wet basis).

The Rhéoplast combines the features of a Couette rotational system (to execute
the melting and extrusion simulating) and a capillary viscometer (to measure the viscosity) (Figure 2). Only the first part of this equipment was used in this experiment.
Starch granules are drawn from the hopper (1) into a cylindrical shearing chamber
(5) by the motion of the annular piston (2). Enzyme (2 µl/g) is added by a syringe.
The inner piston (3) begins to rotate and creates a shear field in the gap between the
barrel and the piston. Here the starch is melted both by mechanical shearing and heat
conduction from the thermostated barrel (4). After a predetermined time, the rotation
is stopped, the piston is moved up to open a clearance at the bottom of the shearing
chamber. After the thermomechanical treatment, the processed starch is pushed into
the injection pot (6) and is ousted, due to gravity and the pressure imposed by the inner piston, as no capillary was mounted during these experiments. The processed
sample is collected at the exit in liquid nitrogen and stored at -20°C until subsequent
analysis.

The shear rate is determined by the rotational rate of the inner piston. The shear rate
can be considered as quasi-uniform because of the thickness of the melting zone (barrel diameter, 21 mm; inner piston diameter, 16 mm; melting zone thickness, 2.5 mm;
melting zone length, 6 cm). For this study the rotation rates were 200, 400 or 700

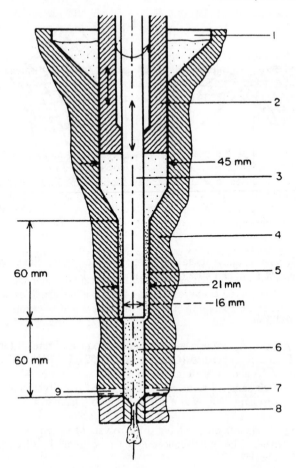

Figure 2   Cross section of the Rhéoplast (Vergnes et al.[3])

rpm, which corresponds to mean shear rates betweeen 75 and 250 s⁻¹. The melt temperature is controlled by the circulation of a thermostated fluid through channels around the barrel. It is measured by a thermocouple (9) located in the injection pot. The temperature was set at 95, 115, 120, 125 and 130°C. The shearing time is the time for which the inner piston is rotated: for this study shearing times were 10, 20 and 30 s.

Analytical Methods.   Enzyme activity was followed spectrophotometrically, by analyzing five samples ($\approx$ 50 mg) of the processed starch, according to the classical method of Nelson.

Enzyme efficiency corresponds to the activity expressed during the thermomechanical treatment. After storage of samples at -20°C, samples were dissolved in 0.1 M KOH, leading to the value R_1 of reducing power. The reducing power R_0, corresponding to blank sample, was measured by dissolving directly native starch and enzyme in a solution 0.1 M KOH, in a mass ratio corresponding to that present during thermomechanical treatment. Enzyme activity inside the Rhéoplast was calculated from R_1-R_0.

Residual enzyme activity after thermomechanical treatment corresponds to the activity measured during 15 min. Samples were ground in liquid nitrogen and then dispersed in boiling water for 5, 10 and 15 min. Reducing power was measured as a function of time, leading directly to residual enzyme activity.

Results

When tested in a 10% starch paste, this enzyme presents a specific activity of 1.278×10^{-4} kat/ml at 95°C. As the enzyme is working in an excess of substrate, its activity has been supposed to be linear, and theoretical amounts of hydrolyzed bonds were calculated for 10, 20 and 30 s of process. The main advantage of the Rhéoplast is that it is able to submit the material to a well characterized thermomechanical treatment on highly viscous samples.

The first set of experiments was carried out on 40% of moisture content (MC) (total wet basis) at a temperature of 95°C. Table 1 shows the effect of shearing time and shear rate on the specific activity of enzyme measured under shear. Values represent the ratio of observed activities to the corresponding activities without shearing, calculated from the specific activity measured on a 10% solution. These data correspond to mean values calculated over the different periods 10, 20 and 30 s.

Table 1   Enzyme activity in the Rhéoplast at 95°C and MC=40%

Rotation rate (rpm)	Shearing time (s)		
	10	20	30
	Enzyme activity (%)		
200	ND*	57.2	ND
400	46.1	49.0	29.3
700	ND	34.9	ND

* not determined

For a given shearing time, there is a decrease of enzyme activity as a function of shear rate. Similarly, when increasing the shearing time, the enzyme activity decreases. From these data, it is possible to calculate the enzyme activity per period of 10 s, since the values given here are mean values on the time interval considered. So, by subtracting the previous periods of 10 or 20 s, it can be seen that, for 400 rpm, the enzyme presents 46% of its activity in solution during the first 10 s, 49% between 10 and 20 s; but its activity decreases to ≈ 0% during the last period of 10 s. - Residual activities (%) were calculated relative to the initial activity (Table 2).

These values do not correspond to the activities measured under shear. For 10 s, the enzyme presents a higher activity after shear treatment than during thermomechanical process: this would be due to the native semicrystalline organization of starch granules which would have limited enzyme efficiency when starting. Once a

Table 2   Residual enzyme activity after treatment (T=95°C, MC=40%)

Rotation rate (rpm)	Shearing time (s)		
	10	20	30
	Residual enzyme activity (%)		
200	ND	21.1	ND
400	77.2	11.3	2.1
700	ND	2.7	ND

molten state is reached, the enzyme presents low values of activity.These results would be interpreted as a consequence of an irreversible inactivation, with no recovery of enzyme functionally, due to the thermomechanical treatment.

The second set of experiments was carried out on 50% wet matter basis, at a temperature of 95°C. Table 3 shows the effect of shearing time and shear rate on the specific activity of enzyme.

Table 3   Enzyme activity in the Rhéoplast at 95°C and MC=50%

Rotation rate (rpm)	Shearing time (s)		
	10	20	30
	Enzyme activity (%)		
200	ND	61.2	ND
400	61.7	55.0	48.3
700	ND	17.9	ND

It can be seen that the activity expressed in the Rhéoplast is greater than for MC = 40%, which can be explained by the lower viscosity of the melt, inducing lower shear forces, but there is still a drop of activity at high shear rates (700rpm). Residual activities were calculated relative to the initial specific activity (Table 4).

Table 4   Residual enzyme activity after treatment (T=95°C, MC=50%)

Rotation rate (rpm)	Shearing time (s)		
	10	20	30
	Residual enzyme activity (%)		
200	ND	6.6	ND
400	91.2	41.4	6.2
700	ND	11.1	ND

Similar conclusions can be drawn from these data. Shear denaturation of enzyme is less drastic in these moisture conditions, but there is still an irreversible loss of activity for longer times and higher shear rates.

The last factor studied is the temperature. In the 50% moisture conditions, the effect of temperature was followed at 400 rpm and 20 s of treatment. Surprisingly, enzyme activity increased up to 120°C, whereupon low reproducible results are obtained (Figure 3). The thermal stability of Thermamyl would be explained by the high starch concentration, which will protect the enzyme from thermal denaturation. The activity of the enzyme is all the more important as starch is encountered in a molten phase.

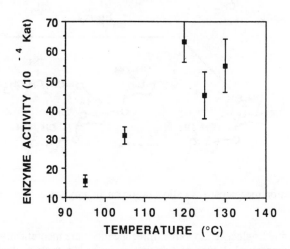

Figure 3   Influence of temperature on enzyme activity expressed in Rhéoplast

This study demonstrates clearly that enzyme use in extrusion process is limited not only by high temperature but mainly for high shear, when increasing the residence time. However high solids concentration would exert a protective effect during the initiation stage, concomitant to starch melting. When residence time becomes higher than 20 s, lower moisture content enhances enzyme sensitivity, probably because of the high viscosity of the medium .

3   CHEMICAL MODIFICATION

The aim of this part was to examine the possible use of a twin screw extruder as a chemical reactor for starch chemical modification, as far as possible, on an homogeneous molten phase, resulting from a melting in a first stage of the extruder. Starch cationization was chosen as a model reaction since it is of great interest for the paper industry[4,5]. The starch is usually slurried in water at a dry weight (d.b.) from about 10 to 42%. Reaction times are generally 10 to 20 hours, at a pH of 11-12; the temperature must be low enough to prevent gelatinization during the process. Necessary adaptations were brought to the extrusion system to give feasibility to the process.

The reaction of cationization involves two stages, a first one for activating the reagent in alkaline medium into an active epoxy form, and, a second one, to operate the substitution on starch backbone (Figure 4).

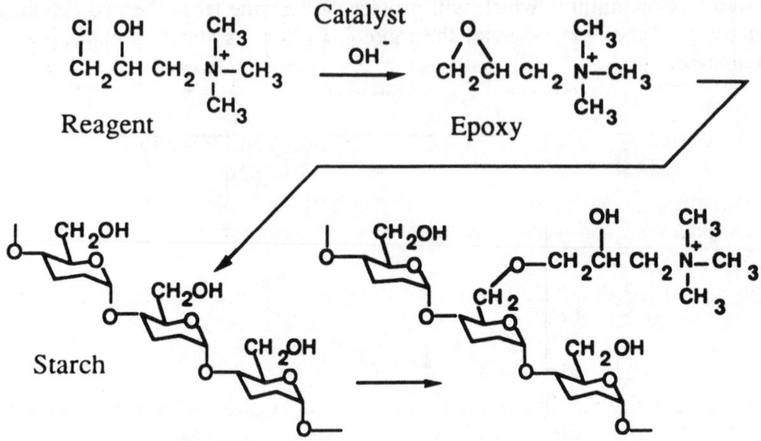

Figure 4   Reaction mechanism of cationization on an amylose chain

For each anhydroglucose monomer, three locations are theoretically available, so the maximum theoretical degree of substitution is 3. The reaction has been extensively studied by Carr and Bagby[6], on a batch process with an excess of water, and their results will be further used for comparison purposes.

Material and Methods

Wheat starch (initial moisture content = 12,5% (wb)) was provided by Ogilvie Aquitaine (F-33000 Bordeaux). The reactive CHPTMA (3-chloro 2-hydroxypropyl-trimethyl ammonium chloride) was purchased from Shell-Chimie (F-78000 Rueil-Malmaison) as a solution Reagent-S-CFZ containing 65% of active CHPTMA (density=1,16 kg/l). NaOH was purchased as anhydrous pellets and mixed in distilled water in order to obtain the desired normality.

Extrusion Equipment.   A pilot scale twin-screw extruder Clextral BC45 was used, with five thermally regulated sections. Its length was 1.25 m and it was terminated by two cylindrical dies ($\emptyset$=4 mm, l=30 mm). A special section for liquid reagents injection was added at the middle of the barrel and the screw geometry featured a first reverse screw element (RSE) before this section and a second one located just before the die (Figure 5).

Water and reagents (CHPTMA as well as NaOH) were fed by two volumetric pumps at the desired location of the barrel. Total throughput was measured by weighing samples at the die outlet for a constant amount of time (1min). Classical measurements were made to assess extrusion treatment, i.e. product temperature at the die and SME (Specific Mechanical Energy Input) by means of the electric consumption of the main motor drive.

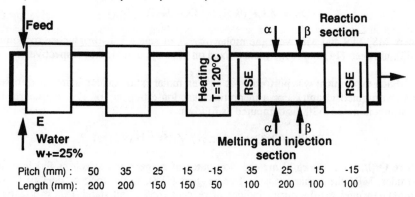

Feed — Reaction section — α ↓ β

Heating T=120°C | RSE | RSE

E
Water
w+=25%

α↓ ↑β
Melting and injection section

| Pitch (mm) : | 50 | 35 | 25 | 15 | -15 | 35 | 25 | 15 | -15 |
| Length (mm): | 200 | 200 | 150 | 150 | 50 | 100 | 200 | 100 | 100 |

<u>Figure 5</u> Experimental device used for starch cationization including screw configuration

Average residence time in the reaction section, i.e., after the first RSE was computed thanks to a software for twin screw extrusion simulation[7].

<u>Extrusion Experiments</u>.   The following variables were changed, in the range indicated:

- screw speed (200-250 rpm), feed rate (20-26 kg/h) in order to change the SME supplied to the product ,
- temperature of the last section of the barrel before the die (50-120°C) in order to change the reaction temperature,
- molar ratio CHPTMA/Starch expressed in anhydroglucose unit (0,055-0,14) by changing both flow rates of CHPTMA and NaOH.

NaOH was injected after the first RSE (position β) and reagent was injected after the first RSE (position α). Other variables were kept constant, i.e. added water w+=25%, barrel temperature in melting section = 120°C. Molar ratio catalyst/reagent was kept at its optimum value 1.75.

<u>Determination of the Degree of Substitution and Reaction Efficiency</u>.   Samples were taken at the die outlet and stored at room temperature in dark lightproof plastic bags. After the desired duration of storage had been achieved, they were ground on a 1mm mesh. For sample corresponding to nil storage (time 0), 150 ml of 0,1 N HCl was added to the sample and a suspension was obtained by wet grinding with a Poly- then precipitated in ethanol for 24h at 4°C, then centrifuged, at pH < 5, in order to stop the reaction quickly. The ground sample was centrifuged for 10 min at 5000 g. After discarding the supernatant, the operation was repeated twice in order to remove the remaining free reagent. After the third centrifugation, the solid sediment was dried in a vacuum oven at 40°C overnight.

Nitrogen content of each sample was determined by Kjeldahl method; the amount of nitrogen in native starch sample was withdrawn and the final quantity noted N(%). The formula for calculating the degree of substitution is the following:

$$DS = MS \bullet (\%N) / (MN - (\%N) \bullet MR)$$

where MS, MN and MR were the molar masses of starch anhydroglucose monomer (162), nitrogen (14) and reagent once fixed on glucosyl unit (152.5) respectively.

Each experiment was performed at a given molar ratio reagent /starch (expressed as anhydroglucosyl unit). For a theoretical reaction efficiency of 100, the degree of substitution, noted DSth, is computed as follows:

$$DSth = (Qr/Qs) \bullet (Ipr. \, r_r / MRf) \bullet (MS / Ips)$$

where $Qr(l/h)$ and $Qs(kg/h)$ are the flow rates of reagent and starch as fed into the extruder, MRf the molar mass of free reagent (188), r_r the reagent density (1160 kg/m^3) , Ipr and Ips the purity indices of reagent (65%) and starch (86.5%) respectively.

Then the reaction efficiency was logically defined by: RE = DS / DSth.

Results and Discussion

First results[8] obtained when performing this reaction either with native starch and extruded starch by simply mixing starch with reagent and catalyst in a conventional twin screw extruder underlined the necessity of achieving the reaction under a homogeneous molten phase. It was thus expected that native granular starch, once it had flowed out of the first RSE, would be under an homogeneous molten phase and that the reaction could be performed in a second stage under a controlled temperature different from that in the melting section. The basic hypothesis relies on our own experience of extrusion process as well as on theoretical computation. In such conditions it was then worthwhile to analyze the influence of reaction parameters as did Carr and Bagby[6].

All the results are given in Table 5. The substitution still went on during storage but could be considered to have reached the plateau after 24 h, so the values of DS and reaction efficiency RE were only measured at that time.

The special screw design allowed us to perform the chemical cationization reaction in thermal conditions different from the temperature reached during the melting stage (> 120°C) (trials 1 to 5). Thus an optimum reaction efficiency (67% for trial 8) was found around 80°C (Table 5). The increase of SME was obtained by an increase of screw speed and decrease of feed rate (trials 1 and 6 to 8). Cationization efficiency then significantly increased from 60 to 69% when SME increased from 209 to 290 kWh/t (Table 5). However, the analysis of the influence of temperature has to take into account the residence time in the reaction section. For this purpose, we have computed the average residence time in the reaction section (t_r) and in the whole extruder (t_e), assuming that only feed rate and screw speed modify this variable. Results are reported in Table 6.

Except for no. 6, t_r and t_e followed the same trend. By comparing both Tables 5 and 6, it can be observed that when t_r decreased, RE increased (Figure 6).

Table 5   Experimental conditions and results of starch cationization by extrusion

Trial no.	Feed rate (kg/h)	Screw speed (rpm)	Reaction temp. (°C)	Inject. pos.	Reagent Starch	Die temp. (°C)	SME (kWh/t)	DS	RE (%)
1	25	200	75	α	0.055	101	209	0.040	60
2	12.5	200	120	α	0.055	150	214	0.015	27
3	25	200	120	α	0.055	127	164	0.026	47
4	25	200	75	α	0.055	80	211	0.037	67
5	25	200	50	α	0.055	55	200	0.032	58
6	25	230	75	α	0.055	90	235	0.035	63
7	25	250	75	α	0.055	102	265	0.037	68
8	20	250	75	α	0.055	103	290	0.038	69
9	25	200	75	E	0.055	100	210	0.039	71
10	25	200	75	E	0.055	89	198	0.040	72
11	25	200	75	E	0.086	88	204	0.060	70
12	25	200	75	E	0.115	90	203	0.078	68
13	25	200	75	E	0.141	92	217	0.090	64

Table 6   Computed average residence times in the extruder

Residence time		Trial no.			
	1	2-3-4-5	6	7	8
t_r (s)	90	38	43	33	25
t_e (s)	183	78	130	120	78

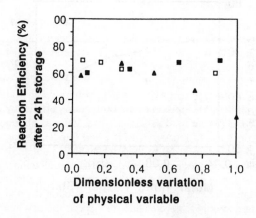

Figure 6   Reaction efficiency as a function of temperature (50-150°C, Δ), computed residence time t_r (20-100 s., □) and SME (200-300 kWh/t, ■)

These results are in good agreement with those of Meuser et al.[9], although the chemical reaction, in that case, was performed in heterogeneous phase leading to an overlapping between starch melting and starch chemical modification. Response surfaces showed that when temperature was above 120°C, for an average SME of 155 kWh/t, RE decreased approximately from 70 to 50% as the total average residence time (t_e) increased from 40 s to 100 s. Moreover, for high SME (> 200 kWh/t), and high temperature (> 135°C), large residence time (> 60 s) led to sharp decrease of RE (~27%) which can be compared to the result of trial no.6.

These facts can be interpreted in two ways: 1) the epoxy form is sensitive to high temperature and degraded as the time of reaction increased and 2) when t_r decreased SME increased and thus starch fragmentation increased leading to a greater availability of macromolecules for substitution due to the chain splitting phenomenon. However, the fact that temperature might increase when SME increases, because of viscous dissipation, shows that these interpretations need further separate experiments for confirmation. Such experiments should involve the use of specific experimental tools able to implement separate variations of physical variables in well controlled conditions, such as the Rhéoplast, used in the preceding part.

The most outstanding improvement was made by injecting the reagent at the entrance E before starch melting, as shown by the comparison between trials 9 and 1, performed under the same conditions of temperature (100°C) and SME (210 kWh/t). This gain (+12%) might be due to a better mixing between reagent and starch in two successive reverse screw elements; after the first RSE, the formation of epoxy takes place in the starch matrix which favor the reaction of substitution. Trials 9 to 13 showed that when the ratio reagent/starch increased, the final DS also increased, from 0.04 to 0.09, but this increase was linked to a slight decrease of the efficiency, from 72 to 64% (Figure 7).

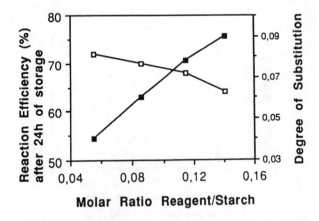

Figure 7   Variations of RE (□) and DS (■) as functions of ratio reagent/starch

In these conditions, the capacity of the extrusion process for manufacturing cationic starches with high degree of substitution was checked. By using the extruder as a two stage reactor, it was thus possible to perform a chemical modification on starch in molten phase. Despite the lack of knowledge on the kinetics of each reaction, these results, in term of reaction efficiency, were quite satisfying since they are comparable to those obtained by heterogeneous extrusion[9] or by the classical batch process[6].

4 CONCLUSION

These experiments have shown that it is possible to perform unconventional use of twin screw extrusion to replace batch processes in the case of chemical modification but not for enzymic conversion. The explanation is probably linked to the specific mechanical effect of extrusion on reagents and substrate. In the case of chemical substitution, shear favors the reaction by increasing the number of sites available by macromolecular chain-splitting. If shear negatively influences the reaction, as in the case of α-amylolysis, due to shear sensitivity of proteins, the use of the extrusion process is questionable. In both cases, the importance of a tool, like Rhéoplast, able to reproduce, in a controlled manner, the same phenomena as those occuring during extrusion has been underlined.

5 REFERENCES

1. P. Linko, 'Extrusion-cooking', AACC, St Paul, 1989, Chapter 8, p. 235.
2. L. Roussel, Thèse de Doctorat de l'Université de Montpellier, 1991, p. 195.
3. B. Vergnes, J.-P. Villemaire, P. Colonna and J. Tayeb, J. Cereal Sci., 1987, 5, 189.
4. D.B. Solarek, 'Modified starches: properties and uses', CRC Press Inc, Boca Raton, Florida, 1986, p. 113.
5. W. Nachtergaele, Starch/Stärke, 1989, 41, 310.
6. M.E. Carr and M.O. Bagby, Starch/Stärke, 1981, 33, 310.
7. G. Della Valle, C. Barrès, J. Plewa, J. Tayeb and B. Vergnes, J. Food Eng., 1992, in press.
8. G. Della Valle, P. Colonna and J. Tayeb, Starch/Stärke, 1991, 43, 300.
9. F. Meuser and N. Gimmler, 'Trends in Food Processing', (Ed. Ang How Ghee), Institute of Food Science and Technology, Singapore, 1989.

The authors wish to thank P. Williams for her excellent technical assistance.

STUDY OF CELLULOLYTIC HYDROLYSIS OF FURFURAL PROCESS WASTES

K. Réczey, E. László and J. Holló[1]

DEPARTMENT OF AGRICULTURAL CHEMICAL TECHNOLOGY,
TECHNICAL UNIVERSITY OF BUDAPEST, P.O.BOX 91
[1]CENTRAL RESEARCH INSTITUTE FOR CHEMISTRY, HUNGARIAN
ACADEMY OF SCIENCES, P.O.BOX 17, 1525 BUDAPEST, HUNGARY

1 INTRODUCTION

In recent years, intensive international research has been conducted to promote wide-ranging utilization of renewable raw materials, such as lignocelluloses, for the production of chemicals, energy, as well as food and feed products. In spite of all efforts, however, no promising results have appeared in the literature on the solution of the problems so far. In subsequent investigations, various economic and engineering problems must also be taken into consideration.

Pretreatment of agricultural and forestry wastes suitable for biotechnological utilization, either by fermentation or by enzymic hydrolysis involves considerable expenses.

Owing to their high energy requirement, physical or mechanical pretreatment, different grinding and cutting procedures cannot be considered as economical technologies either.

Chemical pretreatment is highly efficient, suitable for the preparation of easily hydrolyzable substrates, application of the methods may, however, involve serious problems in neutralizing and depositing alkaline or acidic solutions without causing environmental pollution.

By means of biological and biochemical pretreatment with white rot fungi or with the paradiphenol oxidases, considerable decreases can be attained in the lignin content of various lignocellulose substrates; cellulolytic hydrolysis, however, does not increase satisfactorily.

For the time being, the most suitable method of pretreatment seems to be steam explosion, when, after being preheated to 150°C-180°C, the substance is passed through a special valve and suddenly expanded. With optimal choice of the heat degree and residence time, a satisfactorily pretreated lignocellulose product may be obtained with negligible loss of sugar. Utilization of the product produced by steam explosion as a carbon source for fermentation and as the substrate of enzymic hydrolysis was summarized by San Martin et al.[1], as well as by Dekker[2] and Eklund[3] et al.

Naturally, any pretreatment aimed at the preparation of a useful product, yielding hydrolyzed lignocellulose only as a by-product is also satisfactory. Such technologies are, e.g., furfural production or xylite preparation. In both cases, a by-product similar to that derived by steam explosion is attained, with the only difference that the latter substance contains no hemicelluloses. Another benefit of utilizing these products is that pretreatment, which involves substantial costs, can be eliminated in the enzymic hydrolysis of lignocelluloses.

2 MATERIALS AND METHODS

Description of the Substrate

We studied the enzymic hydrolysis of furfural process wastes obtained from two different sources: on wood base with 45% moisture content and on corn-cob base with 50% moisture content. The outer appearance of the two samples was markedly different: the sample on wood base was fibrous with a loose structure and that on corn-cob base was a fine powdery substance. The particle size distribution is given in Table 1.

Table 1   Particle size distribution of furfural process wastes on wood and corn-cob base

Particle size (mm)	On wood base (%)	On corn-cob base (%)
> 1.4	20.5	10.5
1.0-1.4	10.5	5.0
0.63-1.0	11.5	6.5
0.32-0.63	18.0	15.5
0.20-0.32	11.5	12.3
0.10-0.20	13.0	16.5
0.06-0.10	6.5	20.5
< 0.06	8.5	13.7

The particle size of 50% of furfural process waste on corn-cob base is below 0.2 mm, which is very favourable for enzymic hydrolysis, whereas this particle size could be attained only in 28% with furfural process waste on wood base. Lignin content of the samples was: 46-47%, ash content: 2.4-2.5%, thus, in conversion calculations potential sugar content may be estimated as 50%.

Determination of Reducing Sugar Content.   Sugar content was determined with dinitrosalicylic acid by the method of Miller[4]. For calibration a glucose standard, up to 1 mg/cm^3 was used.

Filter Paper Hydrolyzing Activity (FPU).   FPU was determined according to the IUPAC recommendations compiled by Ghose[5]. Enzyme unit (IU) refers to the number of micromoles of glucose formed from filterpaper-strips (1·6 cm, Whatman Nr.1) in 0.05 M citrate buffer at 50°C, pH 4.8, in a 1 min. reaction time. (Complete hydrolysis time: 1 h, the reaction was stopped and the amount of glucose determined by means of dinitrosalicylic acid.)

Enzymic Hydrolysis.   Experimental conditions for the study of enzymic hydrolysis were:

- substrate concentration (S): 3.3% (3 g dry matter in 90 g total weight)
- enzyme: Substrate ratio E/S: 0.3 g/g = 18.6 IU/gS
- enzyme: NOVO Celluclast 250 S Type N
- pH: 4.8 by application of a 0.05 M trisodium citrate buffer
- temperature (T): 50°C
- reaction time (t): 24 h
- shaking was carried out in a water bath (LaborMIM product).

Enzyme Adsorption and Enzymic Hydrolysis at Various Substrate Concentrations.   Experiments were performed in a water bath (LaborMIM), in shake flasks, with temperatures maintained at 20°C during enzyme adsorption, and at 50°C during subsequent hydrolysis. The amount of substrate was varied between 1 and 5 g, the amount of enzyme added in each experiment was 60 IU (FPU), ensured by 30 cm^3 broth of 2.0 IU/cm^3 activity.

Enzymic Hydrolysis in Two Steps.   Furfural process waste on corn-cob base was used as substrate: 1.3 and 5.0 g furfural process waste (dry matter) was hydrolyzed with 20 cm^3 broth of 2 IU/cm^3 activity each. The first step of hydrolysis lasted 4.5 h; then after removal of the hydrolysate by centrifugation (7000 g) and addition of an identical amount of citrate buffer, hydrolysis was carried on. The second step of the process required 17.5 h.

Enzymic Hydrolysis in Two Steps with Introduction of a Washing Process.   120 g of furfural process waste (50% dry matter content) was hydrolysed with 400 cm^3 of supernatant from Trichoderma broth of 2.0 IU/cm^3 activity. This involved 11.5% substrate concentration and 13.3 IU/gS. The first hydrolysis step took 4 h and the second required 21 h. The remaining sugar was washed out - after the first step - by means of 450 cm^3 citrate buffer (pH: 4.8; 0.05 M).

Cellulase Production in Shake Flasks.   Fermentation experiments were carried out with Trichoderma reesei RUT C30 in 750 cm^3 Erlenmeyer flasks (150 cm^3 useful vol.) with shake revolution no.: 300/min. As inoculum 14-days old konidia we used. Composition of the culture medium was as follows:

Paper wadding	7.5 g
$(NH_4)_2SO_4$	1.4 g
Urea	0.3 g
KH_2PO_4	2.0 g
Yeast extract	0.25 g
Bacto peptone	0.75 g
$MgSO_4 \cdot 7\,H_2O$	0.3 g
$CaCl_2 \cdot 2\,H_2O$	0.3 g
$FeSO_4 \cdot 7\,H_2O$	5.0 mg
$MnSO_4 \cdot 4 \cdot H_2O$	1.6 mg
$ZnSO_4 \cdot 4\,H_2O$	1.4 mg
$CoCl_2 \cdot 6\,H_2O$	2.0 mg

in 1000 ml tap water. After sterilization, pH value ranged between 5.5-5.8.

Cellulase Production on Large Scale. Enzyme fermentation was carried out in a 24 l New Brunswick fermentor, then in a 300 l Brown fermenter (BIOSTAT 300 D) and in a 300 l HTPJ (High Turbulence Plunging Jet) fermenter. As inoculum, 10 vol.% 40-h shake flask culture was applied. The composition of the culture medium was:

Furfural process waste	80.0 g/l
Urea	0.7 g/l
Ammonium hydroxide (20%)	10.7 ml/l
$CaCl_2 \cdot 2 H_2O$	0.8 g/l
$MgSO_4 \cdot 7 H_2O$	0.8 g/l
H_3PO_4 (85%)	2.6 ml/l
Fodder yeast	1.0 g/l
Micromix A trace element solution	0.3 ml/l

3   RESULTS AND DISCUSSION

Hydrolysis of Wet, Air Dried and Hot-Air Dried Furfural Process Wastes

The purpose of our investigations was to determine if despite its high lignin content, furfural process waste can be hydrolysed, and whether there is any difference between process wastes on wood cutting- and/or corn-cob base, and if substrate drying causes any difference in hydrolysis.

Measurements were carried out for furfural process wastes on both wood and corn-cob base, with wet, air dried and hot-air dried (105°C) samples. Results were expressed as reducing sugar concentrations as a function of hydrolysis time and are summarized in Tables 2 and 3.

Table 2   Enzymic hydrolysis of furfural process wastes on wood base, expressed as reducing sugar content as a function of hydrolysis time

Hydrolysis time (h)	Reducing sugar content (mg/cm³)		
	Wet	Air dried	Hot-air dried
0	0.3	0.3	0.3
1	6.9	5.4	3.6
2	9.9	-	4.9
3	10.7	7.1	5.9
6	13.3	8.5	6.8
24	15.4	10.9	9.8

The results obtained with the furfural process wastes on wood base regarding both initial hydrolysis rate and glucose concentration attained in 24 h are approx. 10% lower than the data yielded by the substrate on corn-cob base. This may in part be attributed to the difference in particle size.

On the basis of the measurements it has also been stated that air drying and/or drying at 105°C greatly decreases the rate of hydrolysis and final conversion. By analysis of the initial rates of hydrolysis in case of wood-based substrates, the initial rate of hydrolysis decreased was to 77 and/or 50%, depending on whether it was allowed to air dry or was dried at 105°C.

Table 3   Enzymic hydrolysis of furfural process wastes on corn-cob base, expressed as reducing sugar content as a function of hydrolysis time

Hydrolysis time (h)	Reducing sugar content (mg/cm³)		
	Wet	Air dry	Hot-air dried
0	0.8	0.8	0.8
1	8.1	6.8	3.1
2	11.1	8.1	5.1
3	14.1	11.1	5.4
6	15.5	13.6	7.4
24	17.0	14.9	12.9

Even more significant decreases could be observed in the rate of initial hydrolysis with corn-cob substrates, where as a result of drying, the initial rate of hydrolysis dropped to 32% that of the wet sample.

Figures 1a and 1b show the trend of conversion as a function of time for both substrates subjected to the three conditions.

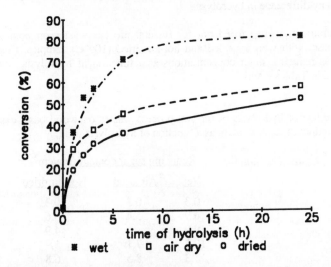

Figure 1a Enzymic hydrolysis of furfural process waste on wood base: trend of conversion as a function of hydrolysis time

Great differences could be observed, especially in the case of the substrate on wood base, between the hydrolysis of wet, air dried and hot-air dried samples (105°C). In the case of the wet sample (50% DMC) in 24h, 84% with the air dried sample 58% and with the sample dried at 105°C, 53.3% conversion could be attained. Air drying and/or hot-air drying decreased the 24-h conversion by 31% and/or 36%. In the case of furfural process waste on corn-cob base, the decrease in conversion due to drying was only 13 and 24%, respectively.

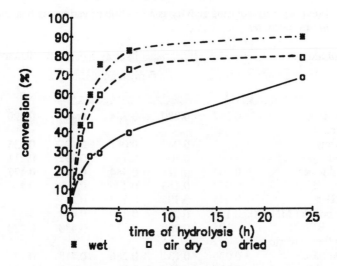

<u>Figure 1b</u> Enzymic hydrolysis of furfural process waste on corn-cob base: trend of conversion as a function of hydrolysis time

In the case of the furfural process waste substrate on wood base, in spite of the slight decrease in initial hydrolysis rate, the decrease in the 24-h conversion was substantial, therefore, utilization in its state of formation (approx. 50% moisture content) is very important, since if it dries out, the micro fibrils will also dry out and will not be able to re-moisturize, which may cause significant decrease in conversion.

<u>Cellulase Production</u>

As low-cost enzyme production is a significant condition for economical biotechnological utilization of lignocellulose. Attention must be focussed on the search for low-cost culture medium components, especially carbon sources. The utilization of furfural process wastes could offer a wonderful potential for this.

Cellulases are inductive enzymes, therefore, in order to attain appropriate level of production with the Trichoderma reesei strain, it was grown on a medium containing cellulose as carbon source. Results attained on different papers, pure celluloses and cellulose-containing wastes in shake flasks are collected in Table 4.

Results similar to those yielded by paper wadding have been obtained on both laminated paper and furfural process waste carbon source. Furfural process waste samples both on wood- and on corn-cob base proved to be good carbon sources, owing to the higher enzyme activities and favourable foaming properties; however, waste on wood base has been examined in detail.

Through a series of culture medium concentrations and by culture medium modifications a culture medium was obtained in which 80 g/l furfural process wastes was applied as carbon source. In the course of scaling up, 24 l laboratory and 300 l pilot-plant fermenters were used. During 110-124 h fermentation 4.2-4.4 IU/cm^3 FPU were attained. In Figure 2 the fermentation conducted in a HTPJ (High Turbulence Plunging Jet) fermenter is presented.

Table 4   Filter paper hydrolysing activity as a function of various carbon sources act-
           ing as substrates

Carbon source	Filter paper hydrolysing activity (IU/cm³)					
	Fermentation time (days)					
	1	2	3	4	7	14
Paper wadding	0.117	0.280	0.384	0.444	0.420	-
Newspaper						
black white	-	0.040	0.081	0.121	0.135	0.151
coloured	-	0.101	0.169	0.192	0.211	0.211
Laminated paper	0.104	0.328	0.484	0.520	0.477	-
Half-cellulose	0.053	0.153	0.396	0.507	0.468	-
Avicel PH10l	0.045	0.105	0.175	0.241	-	0.401
Cellulose powder MN 300	0.115	0.285	0.390	0.444	-	-
Solka Floc BW 40	0.105	0.260	0.375	0.420	0.420	-
Furfural process waste						
on wood base	0.026	0.131	0.266	0.418	0.493	0.505
on corn-cob base	0.026	0.130	0.250	0.402	0.450	0.485
Cornstalk	0.103	0.153	0.206	0.243	0.244	0.248
Sawdust (oak)	0.005	0.005	0.006	0.006	0.006	0.007

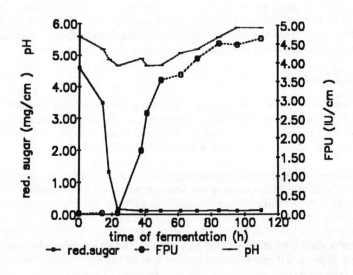

Figure 2   Trend of FPU, reducing sugar content and pH value as a function of fer-
            mentation time in a 300 l HTPJ fermenter

In all further hydrolysis experiments, enzymes prepared in the fermentation experiments were used.

Enzyme Adsorption Investigations

Since furfural production on wood base is hindered by economic problems, while production on corn-cob base seems to be working, we wish now to deal with furfural process waste on corn-cob base in some detail in both adsorption and hydrolysis experiments.

The purpose of our enzyme adsorption experiments was, on the one hand, to determine the highest substrate concentration applicable in a 2.0 IU/cm^3 broth with sufficient amount of enzyme (IU/gS) required for hydrolysis, and on the other hand, to establish whether two-step hydrolysis can be performed without enzyme addition in the second step, using only adsorbed enzyme for hydrolysis. In our investigations, the substrate addition ranged between 0.5-5.0 g, and the amount of enzyme added in each experiment was 60 IU.

Enzyme activity was determined after the elapse of a 50 min. adsorption period.

The results of the investigations carried out with furfural process waste on corn-cob base as substrate are listed in Table 5.

Table 5   Enzyme adsorption and hydrolysis with furfural process waste on corn-cob base with the use of 30 cm^3 broth (2.0 IU/cm^3)

Substrate (g) (%)	Amount of enzyme added to substrate (IU/g)	Amount of enzyme bound to substrate (IU/g)	Conversion in 24-h hydrolysis (%)
0.5 1.6	120	17.4	94.9
1.0 3.1	60	17.0	87.7
2.0 5.9	30	17.5	87.8
3.0 8.3	20	15.9	81.5
4.0 10.5	15	12.2	73.2
5.0 12.5	12	10.1	63.4

According to the above results, independently of the amount of enzyme added to the substrate, the amount of enzyme bound by unit amount of substrate was almost constant in a given range of concentration. For example, at 1.6% substrate concentration the amount of enzyme added was 120 IU/g substrate and at 8.3% substrate concentration enzyme addition was 20 IU/g substrate. The difference in the amount of enzyme bound to the substrate was negligible, showing values of 17.4 and 15.9 IU/g, respectively.

In the experiment with 1.6% substrate concentration, 94.9% conversion was attained, whereas in the subsequent three substrate concentrations conversion markedly decreased, although the amount of enzyme bound hardly changed. In the experiment with 8.3% substrate concentration this was 81.5%, evidently due to product inhibition.

At 10.5 and 12.5% substrate concentrations, with a 2.0 IU/cm^3 broth, enzyme could not be added in the amount required for proper hydrolysis: the 12.2 and/or 10.1

IU/gS enzyme:substrate ratio was not sufficient. The low conversion observed in these experiments may be considered as the joint effect of enzyme deficiency and product inhibition.

Two-step hydrolysis.   For further economical processing of the hydrolysate, higher substrate concentrations and hydrolysates of higher glucose content are required. Therefore, the potentials of multi-step hydrolysis have been examined.

Where no appropriate results could be attained with the one-step process (e.g. the two bottom lines in Table 5), two-step hydrolysis was introduced in order to eliminate at least one of the hindering factors: product inhibition.

Separation of the hydrolysate from the solid waste should be performed after completion of the initial constant rate stage, but still at a point when enzyme adsorption on the substrate is possibly highest.

Our measurement and calculation results as well as glucose concentrations and conversions attained are summarized in Table 6.

Table 6   One- and two-step hydrolysis of furfural process waste on corn-cob base with 20-20 cm^3 broths of 2.0 IU/cm^3 activity

Substrate		Amount of enzyme added	One-step hydrolysis		Two-step hydrolysis		
(g)	(%)	to substrate	Glucose	Conv.	Glucose (mg/cm^3)		Conv.
		(IU/g)	(mg/cm^3)	(%)	Step 1	Step 2	(%)
1	4.5	40.0	24.5	98.0	15.2	13.0	98.2
3	11.5	13.3	49.6	69.1	27.4	45.8	88.8
5	16.6	8.0	51.7	47.0	30.2	57.0	65.0

At 4.5% substrate concentration no difference could be observed between the conversions attained with one- and two-step hydrolyses. At 11.5% substrate concentration, by introduction of the second step, i.e. with one separation process, 28.5% increase in conversion was attained. At 16.6% substrate concentration the improvement was merely 14% since enzyme addition with 2.0 IU/cm^3 broth was not sufficient for this high substrate concentration.

Therefore, in the following experiment, only an 11.5% substrate concentration was studied in an up-scaled system with introduction of a washing step. The complete hydrolysis time was 25 h, from which the first step required 4 h.

As can be seen in Table 7, the first experiment was a one-step process, the second involved two steps, and in the third experiment a washing step was introduced between the two steps of hydrolysis.

In the experiment with 11.5% substrate concentration, 69.4% conversion was attained in one-step hydrolysis during 25 h; in the two-step hydrolysis, conversion was found to be 86.4%. Finally, with introduction of a washing step before the second step of hydrolysis, conversion was as high as 95.1%.

Table 7   One- and two-step hydrolysis of furfural process waste on corn-cob base

Experiment	Step 1		Washing water		Step 2		Conversion (%)
	Glucose concentration (mg/cm^3)	Total glucose (mg)	Glucose concentration (mg/cm^3)	Total-glucose (mg)	Glucose concentration (mg/cm^3)	Total glucose (mg)	
1	49.8	22908	-	-	-	-	69.4
2	32.6	10374	-	-	45.4	18144	86.4
3	32.6	10347	4.15	1867	48.7	19144	95.1

4   CONCLUSIONS

In conclusion it can be pointed out that:

- furfural process wastes on both wood- and corn-cob base can be satisfactorily hydrolysed with cellulase enzymes.

- preliminary drying of the substrate leads to considerable decrease in conversion.

- furfural process waste serves as an excellent carbon source in fermentation for cellulase production.

- cellulase enzyme adsorption on furfural process waste is satisfactory, facilitating multi-step hydrolysis without enzyme addition in the second step.

- considerable increase in conversion can be attained at high substrate concentrations by application of multi-step hydrolysis.

- in conclusion, it may be stated that furfural process waste is highly suitable as a carbon source both in enzyme production by fermentation and as a substrate for enzymic hydrolysis.

5   RFERENCES

1. R. San Martin, H.W. Blanch, C.R. Wilke and A.F, Sciamanna., Biotechnol. Bioeng., 1986, 28, 564.
2. R.F.H. Dekker, Biocatalysis, 1987, 1, 63.
3. R. Eklund, M. Galbe and G. Zacchi, Enzyme Microb. Technol., 1990, 12, 225.
4. G.L. Miller, Anal. Chem., 1959, 31, 426.
5. T.K. Ghose, Pure and Appl. Chem., 1987, 59, 257.

NON-FICKIAN DIFFUSION WITH CHEMICAL REACTION OF A PENETRANT WITH A GLASSY POLYMER: THE GAS-SOLID HYDROXY-ETHYLATION OF POTATO STARCH

N.J.M. Kuipers, E.J. Stamhuis and A.A.C.M. Beenackers

DEPARTMENT OF CHEMICAL ENGINEERING, UNIVERSITY OF
GRONINGEN, NIJENBORGH 4, 9747 AG GRONINGEN, THE NETHERLANDS

ABSTRACT

A mathematical model is presented which describes non-Fickian diffusion with chemical reaction of a penetrant A (ethylene oxide) with some reactive (hydroxyl) group B of a granular glassy polymer (potato starch). The diffusion process cannot be described by Fick's law due to the swelling of the glassy polymer grain, caused by the penetration of the diffusing species. Therefore the kinetics of swelling must be taken into account. A model is presented, allowing for mass transfer with chemical reaction, assuming power-law kinetics for the velocity of the swelling front:

$$\frac{dr_0}{dt} = -K \ (c_{a, \ r=r_0} - c_a^*)^n$$

with c_a^* the threshold concentration for swelling, $c_{a, \ r=r_0}$ the concentration of penetrant A at the position r_0 of the swelling front and K a constant independent of the penetrant concentration. Reaction kinetics of first order in both A (ethylene oxide) and (the reactive hydroxyl groups of potato starch) B (overall second order) is assumed. Criteria for the occurrence of homogeneous addition and for a shrinking core type of reaction are given. Both Fickian diffusion (high K, n > 0 and $c_a^* = 0$) and the so called Case II diffusion (low K, n > 0) are shown to be asymptotic solutions of the model. The assumption of first order kinetics in both ethylene oxide and potato starch is justified by the experimental results. Also the activation energy E_a for the catalyzed gas solid hydroxy-ethylation of potato starch is given for a moisture content of 14 wt% dry basis: $E_a = 45.0$ kJ/mol ethylene oxide (when sodium hydroxide is used as a catalyst).

1 INTRODUCTION

It's known that the diffusion in many polymers cannot be described adequately with Fick's law, especially not when diffusing species cause an extensive swelling of the polymer (Crank[1]). This is particularly the case with glassy polymers. The physical reason for the deviation of Fick's law is the time dependency of the properties of a glassy polymer, due to the finite rate of adjustment of the polymer chains to the presence of the penetrant, often a low-molecular-weight solvent.

Alfrey et al.[2] distinguished three classes of diffusion, based on the relative rates of diffusion and polymer relaxation:

1) case I or Fickian diffusion in which the rate of diffusion is much less than that of relaxation.
2) case II or relaxation-balanced diffusion in which the diffusion is very fast compared to the relaxation processes.
3) non-Fickian or anomalous diffusion which occurs when the diffusion and relaxation rates are comparable.

Case I systems are controlled by the diffusion coefficient. In Case II diffusion the parameter is the constant velocity of an advancing front which forms the boundary between a swollen shell and a glassy core. In anomalous diffusion both the diffusion coefficient and the velocity of the swelling front play a role.

Kuipers and Beenackers[3] reviewed the experimentally observed diffusion anomalies in various polymer systems. The most characteristic features for diffusion in a planar sheet have been obtained at temperatures below the glass transition temperature; there exists a threshold solvent concentration above which the following features are reported:

a) there is a sharp discontinuity in the polymer, which separates a glassy region, where the solvent concentration is negligible, from a swollen rubbery region with a high solvent content.
b) the discontinuity initially moves through the polymer with a constant velocity.
c) initially the amount of solvent absorbed also increases linearly in time.
d) the activation energy for the initial front velocity is not comparable with that for a diffusion process but close to that for craze formation.
e) the front position vs. time curve can be presented by a power-law relation with an exponent ranging between 0.5 and 1.
f) feviations from feature c) occur already before deviations of feature b) are observed.
g) s critical solvent concentration must be reached before Case II diffusion occurs. Also an induction time for Case II diffusion exists: time is needed to create sites in the polymer for the penetrant.

Astarita and Sarti[4] could explain most of the phenomena described above, by taking into account the kinetics of the phase transition. They used a power-law to describe the swelling behaviour of the polymer which takes place at the moving discontinuity and assumed that the rate of swelling depends on how much the local penetrant concentration exceeds some threshold value.

In this paper their model is extended by introducing a chemical reaction of first order kinetics in the diffusing species (i.e. ethylene oxide) and also in the reactive (hydroxyl) group of the polymer (potato starch). Further we adopt the swelling kinetics of Astarita and Sarti[4] while a spherical geometry of the polymer is assumed instead of one-dimensional diffusion in a semi-infinite body as described by these authors. It will be shown that both Fickian behaviour and Case II diffusion are asymptotic solutions of our model of diffusion and chemical reaction of a penetrant in a glassy polymer. The gas-solid hydroxy-ethylation of starch, below its glass transition temperature, is an example of such a non-Fickian process (van Warners et al.[5]); this gas-solid reaction only takes place when a catalyst (sodium hydroxide) is impregnated in the granular potato starch. When this catalyst is not present only absorption of ethylene oxide occurs by the starch granules.

2 THE MODEL

A glassy polymer grain (i.e. potato starch) with radius R, in which absorption of a penetrant A (i.e. ethylene oxide) causes swelling of the polymer, can be schematically presented as shown in Figure 1: the grain consists of a swollen shell (due to the plasticizing effect of A) and a glassy core. The time dependent position of the moving front between rubbery and glassy polymer is r_0. It is assumed that the increase of the diameter of the grain, caused by the swelling, is negligible.

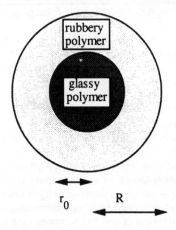

Figure 1   Schematic diagram of the glassy core and rubbery shell for a swelling promoting penetrant diffusing into a grain of a glassy polymer

Assuming a chemical reaction between the diffusing species A and some reactive group B of the polymer we can write:

$$A + B \rightarrow P$$

In case of the gas solid hydroxy-ethylation of potato starch A is the abbreviation of ethylene oxide, B the reactive hydroxyl groups of potato starch and P the product hydroxyl ethyl potato starch. We assume that the diffusing species (ethylene oxide) react according to a (1,1) reaction with reaction rate: $R_a = R_b = -k_{1,1} c_a\, c_b$. In the section with the experimental results it will be shown that this assumption is correct.

If the convective flux of penetrant A in the polymer can be neglected, the unsteady, isothermal mass balance of A in the swollen part of the polymer grain is:

$$\frac{\partial c_a}{\partial t} = \frac{1}{r^2} \frac{\partial}{\partial r}\left[r^2\, \mathrm{ID}_a\, \frac{\partial c_a}{\partial r} \right] + R_a \qquad\qquad r_0 \leq r \leq R \qquad\qquad (1)$$

with c_a the concentration of A (ethylene oxide) in the swollen polymer (potato starch) at radius r at time t , and $\mathbb{D}_a$ the diffusion coefficient of A in the rubbery part of the polymer.

In case of a (1,1) reaction the partial differential equation for reactive (hydroxyl) group B must be solved simultaneously to find the concentration profile of A:

$$\frac{\partial c_b}{\partial t} = R_b = -k_{1,1} c_a c_b \qquad\qquad r_0 \leq r \leq R \qquad\qquad (2)$$

with c_b the concentration of reactive hydroxyl group B of the polymer.

The boundary conditions of this problem are:

$$r = R, t \geq 0: \quad c_a = c_{a0} \qquad\qquad (3)$$

$$r = r_0(t): \quad \mathbb{D}_a \frac{\partial c_a}{\partial r} = -c_a \frac{dr_0}{dt} \qquad\qquad (4)$$

where c_{a0} is the equilibrium solubility of the solvent in the swollen polymer while r_0 is the time dependent position of the interface between rubbery and glassy polymer. Equation (3) is valid because it can be assumed that the induction period is relatively short: i.e. equilibrium occurs almost instantaneously (Cole and Lee[6]). Equation (4) follows from a mass balance across the discontinuous interface with the assumption that no penetrant enters the glassy polymer: this means that the diffusion coefficient in the glassy core is zero which often is an acceptable assumption (Waywood and Durning[7], Tong et al.[8] and Weisenberger and Koenig[9]).

According to Astarita and Sarti[2] we may generally use a power law for the kinetics of the swelling front:

$$\frac{dr_0}{dt} = -K (c_{a,r=r_0} - c_a^*)^n \qquad\qquad (5)$$

A driving force for swelling is assumed which equals the difference between the penetrant concentration at the front and some threshold concentration c_a^* (see also Lasky et al.[10]) a high penetrant concentration results in more swelling and a higher velocity of the front (Harmon et al.[11]). K is a positive constant which is a function of temperature (Lasky et al.[12]), the content of other plasticizers present, e.g. water in case of biopolymers like potato starch (Klech and Simonelli[13]); n cannot be negative. According to Berens[14], Fickian diffusion is observed when the temperature or the concentration of the penetrant are so low, i.e. $c_a < c_a^*$, that the polymer remains glassy. Fickian behaviour may also be expected when the temperature exceeds the unplasticized glass transition temperature of the polymer as is reported by Storey et al.[15]

The initial conditions for this problem are:

$$t = 0: \quad c_a = 0 \qquad 0 \leq r \leq R$$
$$c_b = c_{b0} \qquad 0 \leq r \leq R \qquad (6)$$
$$r_0 = R$$

with c_{b0} the concentration of the reactive group B before reaction starts.

For the glassy part of the polymer the conditions are:

$$c_a = 0 \qquad\qquad 0 \leq r \leq r_0$$
$$c_b = c_{b0} \qquad\qquad 0 \leq r \leq r_0 \qquad (7)$$

Once the front has reached the center of the grain, all the glassy polymer has disappeared and Fick's law is valid everywhere in the grain. From then on the equation of the moving boundary (5) is useless and boundary condition (4) has to be replaced with:

$$r = 0 \qquad \frac{\partial c_a}{\partial r} = 0 \qquad (8)$$

The amount of penetrant A which is taken up by the polymer grain at time t equals:

$$M_t = \int_{t=0}^{t=t} ID_a \left[\frac{\partial c_a}{\partial r} \right]_{r=R} 4 \pi R^2 dt \qquad (9)$$

The maximum amount of A which can be taken up by the polymer consists of a physical absorption term and a reaction term:

$$M_\infty = \frac{4}{3} \pi R^3 c_{a0} + \frac{4}{3} \pi R^3 c_{b0} = \frac{4}{3} \pi R^3 c_{a0} \left[1 + \frac{c_{b0}}{c_{a0}} \right] \qquad (10)$$

with $c_{b0} = 0$ in case of no reaction, i.e. when no catalyst like sodium hydroxide is used.
In practice M_t is the parameter which has to be measured to validate the model.

The differential equations can be rewritten by introduction of the following dimensionless groups:

$$C_a = \frac{c_a}{c_{a0}} \qquad\qquad C_a^* = \frac{c_a^*}{c_{a0}} \qquad\qquad C_b = \frac{c_b}{c_{b0}} \qquad (11)$$

$$r' = \frac{r}{R} \qquad\qquad r'_0 = \frac{r_0}{R} \qquad (12)$$

$$\theta = \frac{K(c_{a0} - c_a^*)^n t}{R} \qquad\qquad Fo = \frac{\mathbb{D}_a t}{R^2} \tag{13}$$

$$\chi = \frac{\mathbb{D}_a}{K(c_{a0} - c_a^*)^n R} \tag{14}$$

$$\phi_{1,1} = \frac{k_{1,1} c_{b0} R^2}{\mathbb{D}_a} \tag{15}$$

χ, which is a reciprocal Péclet number, equals the ratio of the diffusion velocity and the maximum velocity of the swelling front; for a (1,1)-reaction this relation is: $\phi_{1,1} = 9 \cdot Th_{1,1}^2$. With the introduction of θ the time is made dimensionless by introducing the minimum time necessary for the front to reach the center of the grain: with no diffusion limitation, all the polymer is in the rubbery state at $\theta = 1$. Sometimes the Fourier number Fo is chosen as the dimensionless time particularly for easy comparison with the asymptotic case of Fickian diffusion. However in most cases θ is used because of a favourable scaling compared to Fo. Note that values of θ can be converted into Fo by $Fo = \chi \cdot \theta$.

When a constant diffusion coefficient is assumed the dimensionless equations are:

$$\frac{\partial C_a}{\partial \theta} = \frac{\chi}{r'^2} \frac{\partial}{\partial r} \left[r'^2 \frac{\partial C_a}{\partial r'} \right] - \phi_{1,1} \chi C_a C_b \qquad r'_0 \leq r' \leq 1 \tag{16}$$

$$\frac{\partial C_b}{\partial \theta} = -\phi_{1,1} \chi \frac{c_{a0}}{c_{b0}} C_b C_b \qquad r'_0 \leq r' \leq 1 \tag{17}$$

$$\frac{dr'_0}{d\theta} = -\left[\frac{C_a - C_a^*}{1 - C_a^*} \right]^n_{r'=r_0} \tag{18}$$

with dimensionless boundary conditions:

$$r' = 1: \quad C_a = 1 \tag{19}$$

$$r' = r'_0: \quad \frac{\partial C_a}{\partial r'} = \frac{1}{\chi} \left[\frac{C_a - C_a^*}{1 - C_a^*} \right]^n C_a \tag{20}$$

and initial conditions:

$$\theta = 0: \quad C_a = 0$$
$$C_b = 1 \tag{21}$$
$$r'_0 = 1$$

The amount of A which is absorbed by the polymer grain at time θ or Fo equals:

$$M_\theta = \int_{\theta=0}^{\theta=\theta} \chi \, c_{a0} 4\pi R^3 \left[\frac{\partial C_a}{\partial r'}\right]_{r'=1} d\theta = M_{Fo} = \int_{Fo=0}^{Fo=Fo} c_{a0} \, 4\pi R^3 \left[\frac{\partial C_a}{\partial r'}\right]_{r'=1} dFo \quad (22)$$

For Case II diffusion without reaction the relative mass uptake M_θ / M_∞ of penetrant A by the polymer follows from (van Warners and Beenackers[16]):

$$\frac{M_\theta}{M_\infty} = 3\theta - 3\theta^2 + \theta^3 \quad \text{for } \theta < 1 \quad (23)$$

with $\theta = Fo / \chi$.

For Fickian diffusion without reaction, the relative mass uptake M_{Fo} / M_∞ of penetrant in a spherical grain is given by Crank[1]:

$$\frac{M_{Fo}}{M_\infty} = 1 - \frac{6}{\pi^2} \sum_{n=1}^{n=\infty} \frac{1}{n^2} e^{-n^2\pi^2 Fo} \quad (24)$$

An analytical solution of the set of equations (16) - (21) cannot be obtained: the equations have to be solved numerically. The procedure to solve the set of equations is reported by Kuipers and Beenackers[3]. However it is possible to get an approximate analytical solution by introducing a technique known as the pseudo-steady-state approximation. This is worked out by Kuipers and Beenackers[3].

3   SIMULATION RESULTS NON-FICKIAN DIFFUSION WITHOUT REACTION

The concentration profile of penetrant A in the polymer follows from the solution of the equations (16) - (21) with $\phi_{1,1} = 0$. To facilitate comparison with Fickian behaviour Figure 2 presents the concentration profiles for various values of χ together with the profile for diffusion according to Fick's law in case of Fo $= \chi \cdot \theta = 0.01$, $C_a^* = 0$ and $n = 1$. The graph shows that for low values of χ (this means K $\to \infty$) Fickian behaviour is approached: therefore the case of Fickian diffusion is an asymptotic solution of the model for $\chi \to 0$ and $C_a^* = 0$ for $n > 0$. The other boundary of this model: $\chi \to \infty$ results in Case II diffusion behaviour, where the velocity of the swelling front is constant. Between these two asymptotic solutions the model describes anomalous diffusion. Kuipers and Beenackers[3] have derived a mathematical criterion to predict Case II diffusion and Fickian diffusion: Fickian diffusion occurs when $\chi << 0.05$, n > 0 and $C_a^* = 0$, Case II diffusion may be expected to take place for $\chi >> 0.05$, about $\chi \geq 10$.

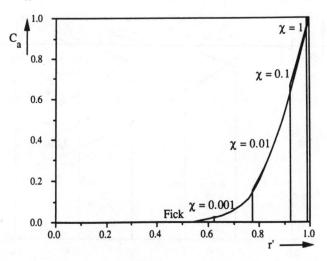

<u>Figure 2</u>   Concentration profiles of penetrant A compared to Fick's solution for various values of χ with Fo = 0.01, $\phi_{1,1} = 1$ and $C_a = 0$

4   SIMULATION RESULTS NON-FICKIAN DIFFUSION WITH (1,1) RE-ACTION

Addition reactions and substitution reactions between a solvent A and some reactive polymer group B can often be described by (1,1) kinetics. This is e.g. the case with the gas / solid hydroxy-ethylation of potato starch (van Warners et al.[5]). In case of a (1,1) reaction the concentration profiles of penetrant A and reactive group B of the polymer are found from the solution of the equations (16) - (21) with $\phi_{1,1} > 0$. Compared to the preceding cases the mass balance of B (equation (17)) has to be solved simultaneously with that of A to find the concentration profile of the latter. Two new dimensionless groups appear in the solution: $\phi_{1,1}$ and c_{a0} / c_{b0}.

Figure 3 presents typical concentration profiles of penetrant A and reactive group B of the polymer as a function of time θ. Always the reaction time is zero at the front; therefore the dimensionless concentration of B equals 1 for $r' \leq r_0'$. The concentration of B is relatively low at the surface of the polymer grain due to the high contact (reaction) time with ethylene oxide.

In case of a (1,1) reaction two asymptotic limits are found with respect to the concentration profile of B (see Kuipers and Beenackers[3]):

a)   a <u>shrinking core type of reaction</u> occurs if:

$$\phi_{1,1} \, \chi \, \frac{c_{a0}}{c_{b0}} > 100 \qquad (25)$$

In this case the reaction almost takes place at the swelling front only. In the rubbery (outer) part of the polymer all B is converted ($C_b = 0$), in the inner glassy core

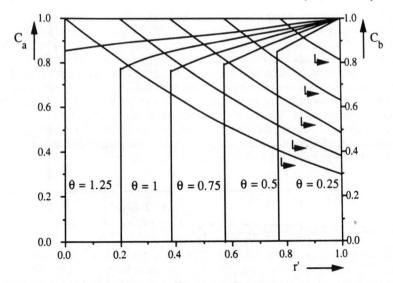

Figure 3   Concentration profiles of A and B for varying times θ with $\chi = 1$, $\phi_{1,1} = 1$, $n = 1$, $C_a = 0.1$ and $c_{a0}/c_{b0} = 1$

of the polymer no B is converted at all ($C_b = 1$). From this equation it is clear that Case II diffusion has a positive influence on a core reaction: the higher χ the more heterogeneous the reaction. This type of reaction may also occur for a high reaction rate $\phi_{1,1}$ and a high concentration of penetrant with respect to the total amount of reactive (hydroxyl) groups initially present.

A special form of a shrinking core type of reaction is surface coating for which the swelling front is located at the surface of the polymer.

b)   Homogeneous conversion of B occurs if:

$$\phi_{1,1} \, \chi \, \frac{c_{a0}}{c_{b0}} > 0.01 \tag{26}$$

Homogeneous conversion of B means the concentration of B has to be independent of the position in the grain, i.e. an uniform degree of substitution occurs throughout the potato starch granule. In case of Fickian diffusion $\chi \rightarrow 0$ and therefore the critical $\phi_{1,1} \, c_{a0} / c_{b0}$ is relatively high compared to anomalous and Case II diffusion: this means that it is easier to get homogeneous conversion of B in case of Fickian diffusion than in case of non-Fickian diffusion. Secondly it can be concluded that not only $\phi_{1,1}$ determines the homogeneity of addition but also c_{a0} / c_{b0} and χ. Equation (26) is only valid for $\phi_{1,1} < 1$. But for $\phi_{1,1} \gg 1$ homogeneous conversion is not possible.

5   EXPERIMENTAL SETUP

In order to determine the absorption and chemical behaviour of the system ethylene oxide and granular potato starch (with a moisture content between 10 and 25 wt% on dry basis), a pressure controlled semi-batch reactor is used. The principle of this reactor, in which the potato starch is contacted with ethylene oxide, is very simple. The pressure of ethylene oxide in the reactor is kept constant, i.e. an amount of ethylene oxide is fed to the reactor which equals the amount of ethylene oxide taken up by the potato starch particles. The ethylene oxide fed to reactor is supplied from a vessel whose weight is measured continuously and registered by the computer as a function of time. After correction of the amount of ethylene oxide present in the gas atmosphere above the starch, the relative mass uptake of ethylene oxide by the starch particles is obtained. Among the most important parameters which are varied in this investigation are the (continuously measured) temperature, the moisture content of the starch and the pressure of ethylene oxide in the gas phase.

6 EXPERIMENTAL RESULTS AND DISCUSSION

<u>Absorption without Reaction</u>

The Figures 4a and 4b show some experimental curves for the relative mass uptake M_t / M_∞ of ethylene oxide by the starch particles as a function of time t for varying values of the ethylene oxide pressure. Temperature and moisture content are relatively low for the conditions presented in Figure 4a: the temperature is 40°C and the moisture content is 9.6 wt% d.b. The polymer is completely in the glassy state as is clear from the fit, using the limit of Case II absorption (equation (5)), which is rather good. The limit of Fickian diffusion is not successful in fitting the experimental measurements. The figure also shows an increasing mass uptake rate for higher values of ethylene oxide pressure in accordance with equation (5) which predicts a higher swelling rate for this case. However more data at different pressures of ethylene oxide are necessary to determine the value of K, n and C_a^* from Figure 4a.

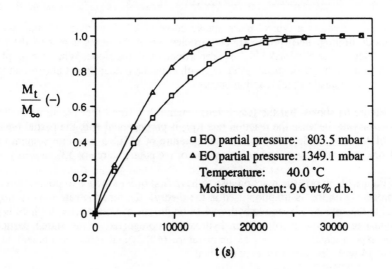

<u>Figure 4a</u>   Relative mass-uptake vs. time at 40°C and 9.6% water content

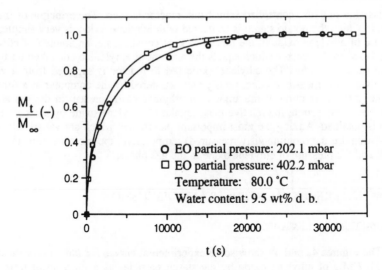

$\underline{\text{Figure 4b}}$ Relative mass-uptake vs. time at 80°C and 9.5% water content

The experimental results shown in Figure 4b are at the same moisture content (9.6 wt% d.b.) but at a much higher temperature of 80°C. Such an increase in temperature results in the creation of more rubbery like starch. This is also made clear by this figure because the experimental results could only be fitted by using the theory of Fickian diffusion. The increase in mass uptake rate with increasing pressure of ethylene oxide is caused by an increasing diffusion coefficient.

Non Fickian diffusion with Chemical Reaction

Figure 5 shows a typical mass uptake curve of ethylene oxide by the (sodium hydroxide impregnated) potato starch particles. From the straight part of the graph the reaction rate is obtained which is proportional with the slope. When the graph becomes straight the particles are saturated with ethylene oxide; no net absorption takes place, only a (steady state) reaction occurs.

Figure 6a shows that the (catalyzed) reaction has first order reaction kinetics in ethylene oxide, because the reaction rate R_{EO} is proportional with the partial pressure p_{EO} of ethylene oxide (at a constant temperature of 60°C, a moisture content of 14 wt% dry basis and a hydroxyl anion content of the potato starch of 132 mol/m³).

Figure 6b reports the reaction also to have first order kinetics in potato starch (in accordance with the assumption used in the theory): the reaction rate R_{EO} is proportional to the amount of reactive hydroxyl anion groups of potato starch (which is proportional to the amount of sodium hydroxide impregnated in the starch particles). Other experimental conditions as temperature (40°C) and moisture content of the starch (14 wt% dry basis) are kept constant.

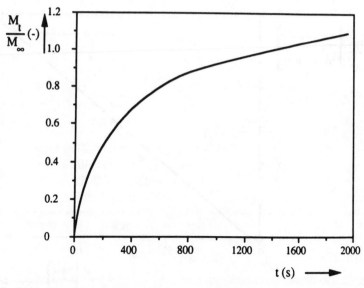

Figure 5 Typical sorption curves of ethylene oxide as a function of time for alkaline starch

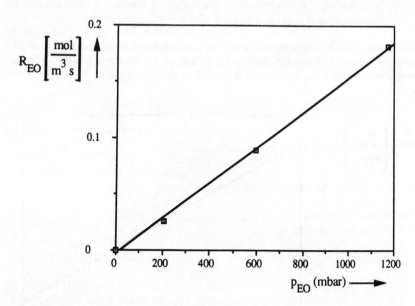

Figure 6a Reaction rate of ethylene function of the partial pressure of this gas

From other experiments it is possible to determine the reaction rate constants $k_{1,1}$ for varying temperature and moisture content of the starch. Figure 7 shows an Arrhenius plot in which the natural logarithm of the ratio of reaction rate and concentration of ethylene oxide in the gas phase (i.e. $R_{EO} / c_{EO(g)}$) is reported as function of

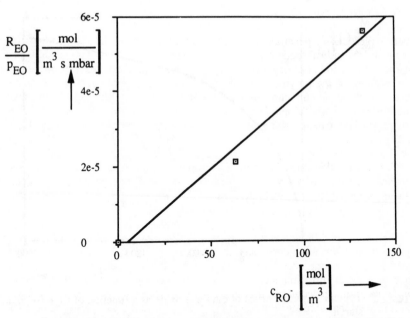

<u>Figure 6b</u> Reaction rate of oxide as ethylene oxide as function of the amount of reactive hydroxyl anion groups of alkaline potato starch

the reciprocal absolute temperature for a moisture content of 14 wt% dry basis. From the negative slope of this graph the activation energy E_a for the catalyzed chemical reaction can be found: $E_a = -$ (slope) R with R as the gas constant. This gives $E_a = 45.0$ kJ/mol.

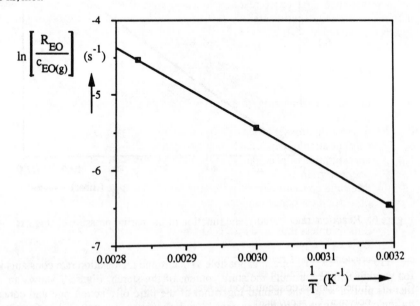

<u>Figure 7</u>   Arrhenius plot for starch with a moisture content of 14%

7   CONCLUSIONS

When a plasticizing penetrant A (ethylene oxide) diffuses into a glassy polymer (potato starch) and subsequently reacts with some reactive (hydroxyl) group B of this polymer a transition of glassy into rubbery polymer is induced, which is accompanied by a swelling of the polymer particles. This has a large influence on the diffusion process of the penetrant into the glassy granular polymer particles. In the present contribution a theoretical and experimental analysis is given of the case of diffusion with power-law kinetics and reaction kinetics of order (1,1) for spherical potato starch particles.

For non-Fickian diffusion without reaction, computer simulations have shown that for high values of χ ($\chi \geq 10$) Case II diffusion occurs, resulting in a constant velocity of the advancing swelling front. Fickian diffusion is another asymptotic solution of the model for $\chi \to 0$, $C_a^* = 0$ and $n > 0$.

Non-Fickian diffusion with a (1,1) reaction has two asymptotic limits: one asymptotic is a shrinking core reaction which is interesting if a surface treatment is aimed at. This regime occurs for $\phi_{1,1} \chi c_{a0} / c_{b0} \geq 100$: this implies that the occurrence of Case II diffusion has a positive influence on a core reaction. The other asymptotic, homogeneous reaction, occurs if $\phi_{1,1} \chi c_{a0} / c_{b0} < 0.01$. From this relation it is clear that homogeneous conversion throughout the particles is more difficult to obtain if Case II diffusion occurs.

For the case of absorption without reaction it is found experimentally that at a relatively low temperature of 40°C and moisture contents of 9.6 wt% d.b. Case II diffusion occurs. At a temperature of 80°C and the same moisture content Fickian theory is necessary for a satisfactory fitting of the mass uptake curves.

The assumed first order reaction kinetics in both ethylene oxide and potato starch is confirmed experimentally. The activation energy for the catalyzed gas solid hydroxy-ethylation is 45.0 kJ/mol at a moisture content of 14 wt% dry basis.

8   NOTATION

A	plasticizing penetrant (ethylene oxide) which diffuses into a (glassy) polymer	[-]
B	reactive (hydroxyl) group of a (glassy) polymer (potato starch)	[-]
c_a	concentration of penetrant A in a polymer grain	[mol·m^{-3}]
c_{a0}	concentration of penetrant A at the external surface of a polymer grain.	[mol·m^{-3}]
c_a^*	threshold concentration of penetrant A for Case II diffusion.	[mol·m^{-3}]
C_a	dimensionless concentration of penetrant A in a polymer grain; $C_a = c_a / c_{a0}$	[-]
C_a^*	dimensionless threshold concentration of penetrant A: $C_a^* = c_a^* / c_{a0}$	[-]
c_b	concentration of reactive group B in a polymer	[mol·m^{-3}]
c_{b0}	initial concentration of reactive group B of a polymer	[mol·m^{-3}]
C_b	dimensionless concentration of reactive group B of a polymer: $C_b = c_b / c_{b0}$	[-]
$c_{EO(g)}$	concentration of ethylene oxide in the gas phase	[mol·m^{-3}]

$\mathbb{D}_a$	diffusion coefficient of penetrant A for diffusion into the polymer	$[m^2 \cdot s^{-1}]$
E_a	activation energy for the gas solid hydroxy-ethylation of potato starch	$[kJ \cdot mol^{-1}]$
Fo	Fourier number, $Fo = (\mathbb{D}_a\, t) / R^2$.	$[-]$
K	swelling constant in the power-law equation of a swelling front	$[m^{3n+1} \cdot mol^n \cdot s^{-1}]$
$k_{1,1}$	reaction rate constant for a (1,1) reaction	$[m^3 \cdot mol^{-1} \cdot s^{-1}]$
M_{Fo}	amount absorbed A by the polymer grain at dimen-sionless time Fo	$[mol]$
M_t	amount absorbed A by the polymer grain at time t	$[mol]$
M_θ	amount absorbed A by the polymer grain at dimension-less time θ	$[mol]$
M_∞	maximum amount of A which can be taken up by the polymer grain: $M\infty = 4\,\pi\,R^3\,(c_{a0} + c_{b0})/3$	$[mol]$
n	exponent in the power-law equation of an advancing swelling front: $n \geq 0$.	$[-]$
r	distance coordinate in a polymer grain	$[m]$
R	radius of a polymer grain	$[m]$
R_a	reaction rate of penetrant A	$[mol \cdot m^{-3} \cdot s^{-1}]$
R_{EO}	reaction rate of ethylene oxide	$[mol \cdot m^{-3} \cdot s^{-1}]$
R_b	reaction rate of reactive group B of the polymer	$[mol \cdot m^{-3} \cdot s^{-1}]$
r_0	position of the front between glassy and rubbery polymer	$[m]$
r'	dimensionless distance coordinate in a polymer grain: $r' = r / R$	$[-]$
r_0'	dimensionless position of the front between glassy and rubbery polymer: $r_0' = r_0 / R$	$[-]$
t	time	$[s]$
$Th_{1,1}$	Thiele number for a (1,1) reaction: $Th_{1,1} = (R/3)\,\sqrt{(k_{1,1}\,c_{b0}/\mathbb{D}_a)}$	$[-]$
T	absolute temperature	$[K]$

Greek Symbols

χ	dimensionless parameter which equals the quotient of diffusion rate and the maximum velocity of the moving boundary: $\chi = \mathbb{D}_a /((K\,(c_{a0} - c_a^*)^n\,R)$	$[-]$
$\phi_{1,1}$	proportional with the square of the Thiele modulus for a first order reaction in A and B: $\phi_{1,1} = (k_{1,1}\,R^2/\mathbb{D}_a) = 9\,Th_{1,1}^2$.	$[-]$
θ	dimensionless time: $\theta = K\,(c_{a0} - c_a^*)^n\,t/R$	$[-]$

9 REFERENCES

1. J. Crank, 'The Mathematics of Diffusion', Clarendon Press, Oxford, 1975.
2. A. Alfrey, F.F. Gurnee and W.G. Lloyd, J. Polym. Sci., 1966, C12, 249.
3. N.J.M. Kuipers and A.A.C.M. Beenackers, accepted for publication by Chem. Eng. Sci., 1993.
4. G. Astarita and G.C. Sarti, Polym. Eng. Sci., 1978, 18, no 5, 388.
5. A. van Warners, G. Lammers, E.J. Stamhuis and A.A.C.M. Beenackers, Starch/Stärke, 1990, 42, 427.
6. J. V. Cole and H.H. Lee, J. Electrochem. Soc., 1989, 136, 3872.
7. W.J. Waywood and C.J. Durning, Polym. Eng. Sci., 1987, 27, 1265.

8. H.M. Tong, K.L. Saenger and C.J. Durning, J. Polym. Sci., 1989, 27, 689.
9. L.A. Weisenberger and J.L. Koenig, Macromolecules, 1990, 23, 2445.
10. R.C. Lasky, E.J. Kramer and C.Y. Hui, Polymer, 1988, 29, 673.
11. J.P. Harmon, S. Lee and J.C.M. Li, Polymer, 1988, 29, 1221.
12. R.C. Lasky, E.J. Kramer and C.Y. Hui, Polymer, 1988, 29, 1131.
13. C.M. Klech and A.P. Simonelli, J. Membrane Sci., 1989, 43, 87.
14. A.R. Berens, J. Appl. Polym. Sci, 1989 , 37, 901.
15. R.F. Storey, K.A. Mauritz and B.B. Cole, Macromolecules,1991, 24, 250.
16. A. van Warners and A.A.C.M. Beenackers, submitted for publication, 1992.

CHARACTERISATION OF MALTODEXTRIN GELLING BY LOW-RESOLUTION NMR

F. Schierbaum[1], S. Radosta[2], W. Vorwerg[2], V. P. Yuriev[3], B.B. Braudo[3] and M. L. German[3]

[1] KANTSTRASSE 4, O-1570 POTSDAM, GERMANY
[2] FRAUNHOFER-INSTITUTE FOR APPLIED POLYMERIC RESEARCH TELTOW, O-1505 BERGHOLZ-REHBRÜCKE, GERMANY
[3] INEOS, 125080-MOSCOW, RUSSIA

1 INTRODUCTION

Potato starch derived maltodextrins of low degree of hydrolysis exhibit the outstanding property of forming thermally reversible gels[1]. If concentrations are > 12% (wt/wt) the homogeneous one phasic starch solutions set back into a two- phasic aggregation network[2]. The temperature must be below 40°C. On heating, these gels will liquefy depending on concentration at temperatures > 40°C. The preparation, properties and applications of this special type of maltodextrin have been reviewed in detail[3].

Special attention was given to the processes of reversible sol-gel transition as well as the properties as thermally reversible gels[2,4,5]. The gelling process has been characterised to be low in cooperativity. Small changes in entropy are sufficient for the formation of a stable aggregation network[4]. The initial process seems to consist of binary interactions between α-1,4-glucosidic chains. The velocity of gelling in the initial phase is high as has been shown by measurements of the rise in viscosity and shear modulus, which are further determined by the degree of hydrolysis of the sample, of concentration and of temperature[2]. About 24 h after onset of gelling, about 12-16% of the polysaccharides are found to be arranged in crystalline junction-zones[5.] Their structure is that of the crystalline B-polymorph of amylose. The gels exist as two-phasic systems consisting of these crystalline domains, which are embedded in a "liquid" phase containing the non-crystallizing parts of the maltodextrins. Interactions between the amylose molecules and some exterior, sufficiently long amylopectin chains in forming the gel structure are likely[6,7].

Further characterization of maltodextrin gelling by means of nondestructive measurements with a minimum of measuring time for each state in sol-gel transition is needed. Two questions require to be answered in connection with the present results:

- whether crystallinity is a secondary, longtime consequence or preceding alignments and network arrangements or if it is primarily responsible for initiation of the entire gelation,

- the mutual role of amylose/amylopectin interaction in the network formation in the special case of hydrolyzed starch polysaccharides and in contrast to the findings of incompatibility between amylose and amylopectin in their high molecular state[3,8,9].

The low-resolution NMR-technique was applied to meet these demands.

2 EXPERIMENTAL

NMR-Measurements

The theoretical background of the measurements is in the transition of a homogeneous, one-phasic maltodextrin solution into a two-phasic system. It was reported to consist of a solid-like structured network which immobilises the liquid phase in a more mechanical way.

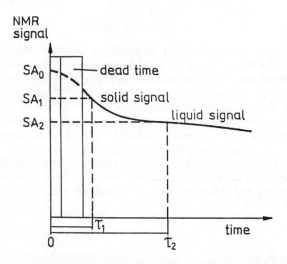

Figure 1 Free induction decay of a radio frequency pulse with time, showing its "solid" and "liquid" parts and the points where the signal amplitude is measured

There are different binding states of the protons which can be discerned by different relaxation times in the low-resolution NMR-technique. If a radio frequency pulse is worked on for a few micro-seconds, all the hydrogen nuclei of the system are excited to rotate by 90° with respect to the static magnetic field[10]. They immediately return to their original state emitting an NMR-signal if the pulse is switched off. The duration of the following "free induction decay" depends on the physical state of the protons; they decay more quickly in the solid phase than they do in the liquid phase. Figure 1 shows, that starting from a constant amount of protons, the entire relaxation plot is composed of two superimposed signals from the solid bonded and from liquid bonded protons. The NMR-signal is measured at two points: at SA1 after 15 μs and at SA2 after 70 μs after the pulse. Regarding a necessary dead-time, the first one is proportional to the total amount of protons in both phases whereas the second one arises from the liquid phase protons only. Regarding instrumental conditions like dead-time and field inhomogeneity (F) as well as digital offset (ϕ), the ratio of the two signals is proportional to the "solid/liquid-ratio" of the system at the moment of the measurement, called "s/l - ratio".

$$s/l = \frac{(SA1 - SA2) \ F \ - 100}{(SA1 - SA2) \ F \ + SA2 + \phi}[\%]$$

The measurements were performed in a BRUKER MINISPEC "pc 120" with an operating frequency of 20 MHz. Results from s/l-working and calibration modus are considered as arbitrary units. Only s/l-values > 0.5 were used for current considerations because lower absolute values are uncertain (region of noise).
The s/l-measurements were carried out using

> concentration 15...35% wt/wt,
> temperature   4°C, 20°C, 25°C,
> time   until reaching equilibrium s/l-values.

The test tubes containing the freshly prepared and instantly cooled test solutions were stored in the thermostat at the desired temperature. They were transferred to the probehead of the device for measurement. For each measurement 4 s were needed, so that the temperature could be kept constant. After this the tubes were returned to the thermostat. For measurement of the relaxation time T2 as well as for solid/liquid relation P2 the polysaccharides (amylose, amylopectin) were dissolved in sodium hydroxide solution and neutralized (pH 6.3, final concentration 1 M NaCl). The solutions were heated to 95°C at the end of neutralization, transferred to the test tube, stoppered and positioned in the thermostat at the onset temperature of 80°C. Measurements were performed during cooling to 22.5°C and final holding at 25°C. The cooling period was run within 6 h, additional controls were taken after 24 h and 48 h.

<u>Materials</u>

The maltodextrin (MD) used was a technical sample of SHP, produced for application in the food industry[11] by Stärkefabrik Kyritz, Germany (Table 1).

<u>Table 1</u>   Molecular parameters and average saccharide composition of a maltodextrin SHP

Dextrose equivalent:	6.2
$\overline{DP}_n$, calculated from reduced value:	27
$\overline{DP}_w$, by light scattering in 0.01 M NaOH:	1100
$\overline{DP}_w/\overline{DP}_n$:	41
Iodine complexing value:	λ_{max} 548 nm, E 540 0.110
Oligosaccharides (soluble in 80% ethanol):	18 + 3%, $\overline{DP}_n$ 7
Polysaccharides (insoluble in 80% ethanol):	82 + 3%, $\overline{DP}_n$ 70
Linear polysaccharides (cyclohexanol precipitate):	8 - 11%, $\overline{DP}_n$ 50 - 150

Acetyl-maltodextrin (Ac-MD), laboratory sample, prepared from acetylated potato starch[12]: DE-value 6.3%, acetyl-content 5.2%, DP_n 26, λ_{max} 521 nm (10^{-4} N iodine solution).

Amylose, purified sample of SERVA-amylose JBV 20%,
      hydrolyzed sample of SERVA-amylose $\overline{DP}_n$ 36.

Amylopectin, commercial sample from AVEBE, Veendam, The Netherlands, purified
by removing the linear components[13].

3 RESULTS AND DISCUSSION

NMR-response of MD-gelling

The gelling process of the MD-solution is reflected by rising s/l-values from the
first few minutes until reaching constant values at the equilibrium after some days
(Figure 2). The onset is followed by a period of high velocity development of the s/l-
values. This will be illustrated best by the time for reaching the half of the final
(equilibrium) s/l-values (Table 2).

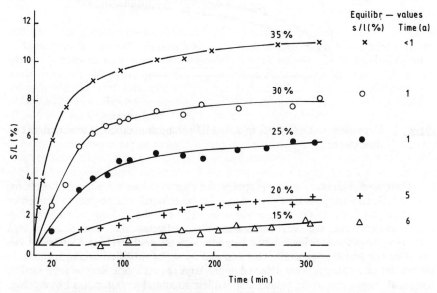

<u>Figure 2</u> Development of the s/l-values of MD-solutions during the first 6 h after
onset of gelling

<u>Table 2</u> Half-value times of MD-structuring varying with concentration (temp. 4°C)

Conc.	s/l-values (%)		Half-value time
(%)	final	half-value	(min)
35	12	6	20
30	9	4.5	40
25	7	3.5	60
20	5	2.5	180
15	3	1.5	300

Gels are formed visibly within this first period of structuring at all concentrations. The exact gelpoints, however, cannot be reflected by this method of measuring. The initial phase will be followed by a phase of decreasing velocity (Figure 3). Equilibrium was reached depending on temperature and concentration after a minimum of 24 h. In the case of s/l-measurements applied here the concentration of dependence of the final values exhibits linearity for each temperature (Figure 4). Hence, it is possible to approximate to a suggested minimum concentration for the formation of aggregated structures.

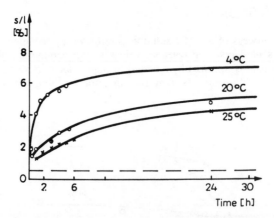

Figure 3   Dependence of s/l-values of 25% MD-solutions at different temperatures on time (30 h)

Compared with other methods applied for characterizing the gelling of the same type of MD, an analogous time dependence has been found with respect to the degree of x-ray crystallinity, to the shear modulus as well as to the enthalpy of melting (Figure 5). In every case the same behaviour will be reflected: the initial phase of high velocity aggregation is followed by a longtime increase until reaching equilibrium values. This last period is generally characterized by slight structural arrangements[14]. In the present case changes were detected until 6 days at maximum. Results submitted by Ring et al.[15], give rise to the assumption that late structural arrangements like crystallization of amylopectin may still be detected after 40 days of storage. Highly ordered structural elements are the essential constituents of the MD-gel networks. A linear relation between s/l-values and average x-ray-crystallinity in the first 60 h of gelling is indicative for this structural basis (Figure 6). Solid structures, however, will be detected already in pre-gel states. From this it follows, that the formation of rigid structural element and separation of phases starts during the early stages of the gelling process. This is in accordance with the findings of Bulpin et al.[16] on 35 wt/wt MD-solutions: structural elements were detected by their storage and excess moduli 16 min after the temperature had dropped below 30°C.

In contrast to the gelling MD-solutions, molten MD-gels, solutions of amylopectin as well as of acetylated MD do not give solid signals; they behave like normal aqueous liquids. This is in accordance with their amorphous scattering in x-ray analysis[17] as well as in high-resolutions NMR[16] and in DSC-melting analysis. On the other hand, MD-liquids in a pre-gel state or below minimal gelling conditions will give s/l-values

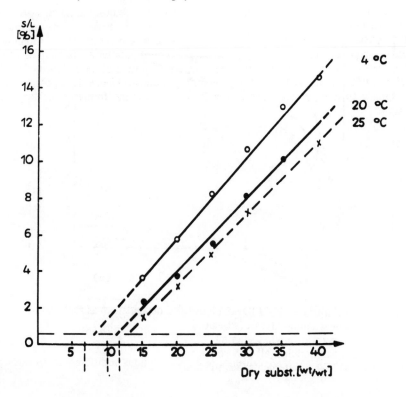

<u>Figure 4</u>   Dependence of equilibrium s/l-values of MD-solution on concentration at 4°C, 20°C, 25°C

and crystalline scattering if aggregation really occurs.This behaviour is independent of reaching a visible gel structure.

Hence it can be stated, that low-resolution NMR reflects the formation of structural elements, which contain a proportion of x-ray crystalline domains. The s/l-values are real findings but up to now an exact calibration against "standard crystallite content" is missing.

<u>Polysaccharide Interaction</u>

Undoubtedly the linear amylose in its dissolved state is responsible for the formation of the x-ray structure in amylose gels as well as in retrogated amyloses. The formation of amylose aggregates from a homogeneous solution (80°C) during its cooling was followed by measuring the relaxation time T2 (ms)[10]. It is shown in Figure 7 that the original T2 signal splits into two signals of different relaxation times, T2a and T2b,when a temperature of 56°C is reached. The former relates to the liquid phase with the longer relaxation times whereas the latter represents the solid part with the shorter relaxation times. By this behaviour the separation of phases and the formation of rigid amylose structures are demonstrated. The T2-values become constant at about

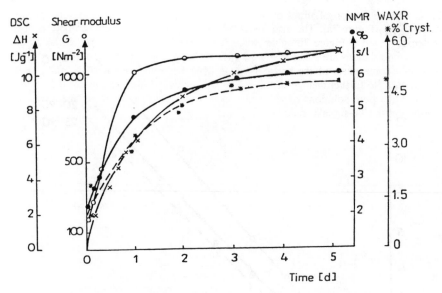

Figure 5   Sol-gel transitions of MD-solutions (time dependence) as revealed by
- •   low resolution NMR (s/l-values)
- *   x-ray wide angle scattering
- O   shear modulus (all 20% solutions)
- x   DSC-measurements (8% solution)

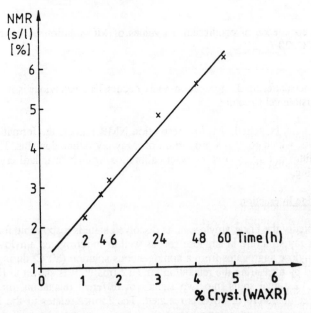

Figure 6   Proportionality between s/l-values and average %-degree of x-ray crystallinity (25%, 20°C)

35°C with relaxation of times of 40 ms and 12 ms for the liquid and solid phases respectively (Figure 7a). The distribution of the polysaccharides between both phases is demonstrated in Figure 7b by the values of P after splitting of the signals. At equilibrium the relations are P2a : P2b = 84 : 16. From this, the conclusion may be drawn that only some parts of the entire amylose have been changed into the solid state within 6 h. No further change of the values was observed after storage of the gel for 24 h. When pure solutions of amylopectin were exposed to the same treatment, no splitting of the T2-signals could be observed.

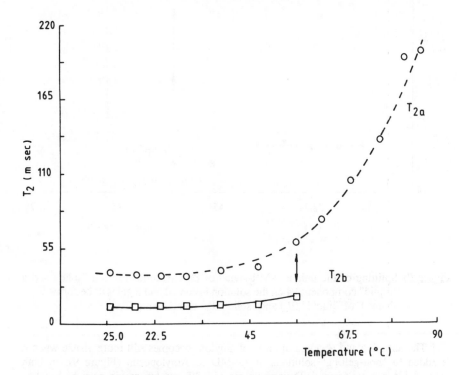

<u>Figure 7a</u> Splitting of the unique NMR-signal of 80°C-solutions into a "solid" and a "liquid" component when the solution is cooled and a gel will be formed relaxation time T_2 - T_{2a} (liquid), T_{2b} (solid) (ms)

From these findings it is clear, that amylose is responsible for the "solid signals" in the case of low-resolution NMR-measurements. The significance for the entire gelling process will be demonstrated by Figure 8. Solutions of amylose were added to gelling MD-solutions at two levels of concentration. The effect is distinctly visible in the initial phase of gelling by lower half-value times, i.e. higher velocity of aggregation. The enhancement of the final s/l-values is more pronounced in the case of the lower concentrated solutions. It may be concluded, that the additional amylose has introduced some more MD-polysaccharides into the gel network. Analogous behaviour has been observed in case of measurements by low-shear rheology as well as by the shear[2] modulus.

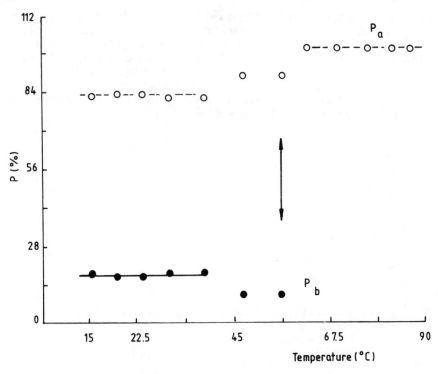

<u>Figure 7b</u> Splitting of the unique NMR-signal of 80°C-solutions into a "solid" and a
"liquid" component when the solution is cooled and a gel will be formed
shares P of "liquid" (P_a) and "solid" (P_b) components

The initiating and accelerating role of amylose becomes still more visible when it
is added to "non-gelling" solutions of Ac-MD or Amylopectin (Figure 9). In both
cases stable gels are formed. Their s/l-values with 2% and 5% amylose are higher than
in the case of pure gels from 6% amylose. As could be expected, retrograded amylose
will also give solid signals (Figure 9), but in this state it is unable to act for initiation or
acceleration of network formation.

These results are supported by DSC-measurements on the same system
(Figure 10). Non-gelled, liquid Ac-MD-solutions exhibit no melting enthalpy, whereas
a gel of amylose has a measurable endothermic peak. When low-molecular amylose
and Ac-MD were exposed to common gelling another melting endotherm was found:
it was characterized by a shift in the melting interval to lower temperatures but to a
higher ΔH-value.

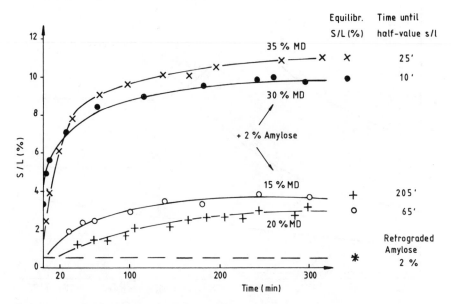

Figure 8   Influence of added amylose on the time dependence and final s/l-values (4°C) of MD-solutions
- O   15% MD-solution + 2% amylose
- +   20% MD-solution without amylose
- ·   30% MD-solution + 2% amylose
- x   35% MD-solution without amylose

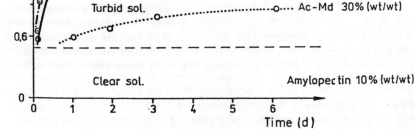

Figure 9   Influence of added amylose on the development of s/l-values of non-gelling solutions of Ac-MD and of amylopectin (20°C)

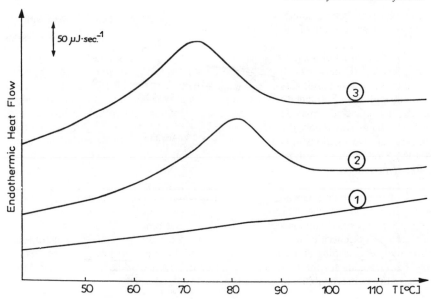

<u>Figure 10</u> DSC-melting endotherms of solutions of amylose, Ac-MD and of a mixture
thereof (calorimeter DAS-4, acc. Privalov, heating rate 1 K/min)

Curve	Sample	Reference	Melting Enthalpy $(J \cdot g^{-1})$	T_A	T_P ($^{\circ}$C)	T_E
1	8% Ac-MD	water	-	-	-	-
2	2% Amylose ($\overline{DP}_n$ 36)	water	12.7	62.2	81.0	93.5
3	8% Ac-MD + 2% Amylose ($\overline{DP}_n$ 36)	8 % Ac-MD	17.0	52.0	72.5	88.0

4   CONCLUSIONS

Experimental proof is given that low-resolution NMR-technique is suitable for following sol-gel transitions if they are based on the formation of a diphasic network structure. The time dependences which are expressed by changes in the solid/liquid relations are quite similar to those of other destructive and non-destructive methods like
low-shear rheology, shear modulus, DSC and x-ray scattering. A linear proportionality
between s/l-values and x-ray crystallinity indicates a close relationship with formation
of real crystalline structural elements as the junction zones within a gel network. These
highly structured domains become detectable from the very onset of gelling after a few
minutes when the system is still in a liquid state. From this observation can be derived
the sequence of processes in maltodextrin sol-gel transition: aggregation of linear sequences, crystallization and separation of phases. At sufficiently high concentrations
the crystalline domains may form a stable reversible gel. However, the gelpoint as well
as the physical state of the system, liquid or solid, will not be reflected by this NMR-

technique. Non gelling solutions, however, may be discerned clearly by s/l-signals of < 0.6% or if no splitting of the T2-signal occurs.

Amylose in its dissolved state is the initiating and accelerating component in the formation of MD-networks. It obviously contributes to the amount of junction zones too and in that function it acts on the other limited or non gelling polysaccharides in a "structuring" way. Sufficiently mobile chains of the amylopectin[6] as well as Ac-substituted glucosidic chains may be inserted into a common gel network. The critical length of these chains was reported to be in the order of 10-12 AGU[18,19]. The results of our rheological investigations[2,7], can be confirmed by the present NMR-investigations: the self-aggregation of the amylose by retrogration will be directed to the formation of a stable hydrated gel.

Our interpretation depends on the assumption of compatibility between the interacting linear and branched polysaccharides. This was strongly denied for the high molecular amyloses and amylopectins[8,9]. This difference between the two interpretations of starch polysaccharide interaction in the gelling process may result from the different molecular basis leading either to thermally reversible or irreversible gels. The solution may be in looking for the molecular and structural scale on which the transition from reversibility to irreversibility in gelling behaviour will take place.

5 SUMMARY

The formation of thermally reversible gels from low-DE potato maltodextrins as well as the structuring of amylose, amylopectin and Ac-maltodextrin was investigated by the pulsed low-resolution NMR-technique. The values were presented preferentially in their "solidus/liquidus relations" as arbitrary units. It could be proved that the applied method is suited for following the sol-gel transitions of the maltodextrins because they are based on the formation of highly ordered (x-ray crystalline) domains. They can be detected after only a few minutes of interactions in the pregel state. Nongelling branched molecules as well as Ac-maltodextrins may be inserted into the gel networks by the structuring action of amylose.

6 NOTES

The experiments were done in the course of collaborative works between the (former) Zentralinstitut für Ernährung (now Deutsches Institut für Ernährungsforschung), Potsdam-Rehbrücke and the Institute for Organo-Element Compounds, of the Russian Academy of Sciences, Moscow.

The contents of this lecture has been published already with extended and slightly modified contents by Carbohydrate Polymers 18 (1992) 155 under the title "Formation of Thermally Reversible Maltodextrin Gels as Revealed by Low Resolution H-NMR".

We are greatly indebted to ELSEVIER SCIENCE PUBLISHERS LTD for the permission of reproducing Figures 1,3,4,6-10 in this paper.

7 REFERENCES

1. M. Richter, F. Schierbaum, S. Augustat and K.-D. Knoch, US-Pat. 3 962 465, 3 986 870 (1976).
2. W. Vorwerg, F. Schierbaum, F. Reuther and B. Kettlitz, 'Biological and Synthetic Networks', (Ed. O. Kramer), Elsevier Appl. Sci., London and New York, 1988, pp. 127.
3. F. Schierbaum, Thesis Dr.sc.nat., Humboldt-Universität, Berlin, 1988, p. 211.
4. F. Schierbaum, F. Reuther, E.E. Braudo, I.G. Plashchina and V.B. Tolstoguzov, Carbohydr. Polym., 1990, 12, 245.
5. Ch. Gernat, F. Reuther, G. Damaschun and F. Schierbaum, Acta Polymerica, 1987, 38, 603.
6. M.J. Gidley and P.V. Bulpin, Carbohydr. Res., 1987, 161, 291.
7. F. Schierbaum, W. Vorwerg, B. Kettlitz and F. Reuther, Nahrung/Food, 1986, 30, 1047.
8. S.G. Ring, Starch/Stärke, 1985, 37, 80.
9. M.T. Kalichevski and S.G. Ring, Carbohydr. Res., 1987, 162, 323.
10. Bruker, MINISPEC application, note 30, 1988.
11. F. Schierbaum, M. Richter, S. Augustat and S. Radosta, Dtsch. Lebensm.-Rundschau, 1977, 73, 390.
12. B. Kettlitz, Thesis Dr.rer.nat., Academy of Sciences, Berlin, 1983, p 163.
13. M. Richter, S. Augustat and F. Schierbaum, 'Ausgewählte Methoden der Stärkechemie', Fachbuchverl., Leipzig, 1964, 2. Auflage.
14. P.D. Orford, S.G. Ring, V. Carrol, M.J. Miles and V.Y. Morris, J. Sci. Food, Agric., 1987, 39, 169.
15. S.G. Ring, P. Colonna, K.J. I'Anson, M.T. Kalichevski, M.J. Miles, V.J. Morris and P.D. Orford, Carbohydr. Res., 1987, 162, 277.
16. P. Bulpin, A.N. Cutler and I.C.M. Dea, 'Gums and Stabilizers for the Food Industry', (Eds. G.O. Phillips et al.), Pergamon Press, Oxford and New York, 1984, pp. 475.
17. F. Reuther, P. Plietz, G. Damschun, M.W. Pürschel, R. Kröber and F. Schierbaum, Colloid and Polymer Sci., 1983, 261, 271.
18. B. Pfannemüller, Int. J. Biol. Macromolek., 1987, 9, 105.
19. M.J. Gidley and V.P. Bulpin, Carbohydr. Res., 1987, 161, 291.

Subject Index

Acetyl-maltodextrin, 280
    composition, 280
    gelling behaviour, 289
Activation energy, 262
ADP-glucose pyrophosphorylase, 36
Ageing, 77, 123, 176
Aggregation network, 278
Agrobacterium, 33
α-Amylase, 19, 28, 134, 137, 233
    active site, 234
    barley, 234
        recombinant barley
            α-amylase, 235
    $(\beta/\alpha)_8$-barrel domain, 233
    Ca^{2+}-binding, 235
    catalytic mechanism, 233
    mutational analysis, 234
    site-directed mutagenesis 237,
    starch binding site, 234
    subsites, 233
    Taka-amylase A, 234
β-Amylase, 16
Amyloplast, 37
Anti-sense RNA, 38
Apparent viscosity, 63
Arabinoxylan, 3, 148
    apple, 3
    wheat, 3
Arepas, 132

Baked products, 63, 210
Barley, 26, 131, 182, 203, 220
Biodegradable plastics, 173
    biodegradability tests
        (Mater-Bi), 173
    foamed, 173
Biopolymer systems, 44
Bird-Carreau model, 43
Biscuit products, 210
Branching enzyme, 37, 233
Bread, 122, 127, 138, 150, 210

Caloric reduction, 164
    fat replacement, 163
    low DE potato maltodextrin, 163,
        289
Capillary rheometer, 64
Carbohydrase action, 5, 234
Carboxypeptidase, 236
Carrageenan, 43, 76, 220
Cassia tora/obtusifolia, 191
Cell wall, 27, 133, 141, 147, 215
    components, 148
        water insoluble, 148, 221
        water soluble, 148
    plant, 31, 33, 124, 147
    polysaccharides, 31, 123, 147,
        215
Cellulase, 4, 27, 204, 254
    cellulolytic hydrolysis, 252
    production, 254
Cellulose, 27, 53, 121, 148, 176,
        199, 205, 215, 255
Ceratonia siliqua, 191
Cereals, 138, 148, 210
Cesalpinia spinosa, 191
Cholesterol, 124, 149. 163, 206
Cluster structure, 18
Coating, 168, 180, 191, 270
Colon, 137, 147, 163
Complex carbohydrates, 125, 163,
    206
Constipation, 206
Constitutive models, 43
Cyamopsis tetragonoloba, 149, 191

β, i, β-Degradation, 16
γ- i-Degradation, 16
Density, 65, 86, 129, 208, 215, 235,
    246
    of cereals, 86
    hysteresis of, 92
    influence on gelatinization, 94

isobars of, 94
isotherms of, 97
of starch, 95
of wheat flour, 95
Desserts, 160
firm textured, 160
milk-based, 160
pourable, 160
ready-to-eat, 160
spoonable, 160
Diabetes mellitus, 147
Dietary fibre, 123, 128, 147, 165,
  205
Dietary guidelines, 121
Diffusion, 128, 150, 222, 235, 262
  case II, 262
  Fick, 262
  non-Fickian, 262
Dilute solution theories, 44
Disaccharides, 121
Doi-Edwards model, 50

E. coli ADP-glucose
              pyrophosphorylase, 37
Empirical constants, 45
Endohydrolase, 26
Enzymes, 31, 34, 104, 127, 138,
              151, 160, 192, 216, 233,
              240, 253
  activity, 36, 138, 242, 257
  amylolytic enzymes, 127, 233
    architecture of, 128, 222, 233
Enzymic conversion and
              fermentation, 240
  of agricultural waste, 252
  of forestry waste, 252
  of furfural process waste, 253
  pretreatment, 252
  steam explosion, 252
Enzymic degradation, 4, 170
Ethylene oxide, 196, 262
Extrusion process, 100, 240
  cationization by, 245
  enzymic conversion by, 240
  melting, 240
  plastification, 175
  reaction efficiency, 247
  residence time, 245
  thermomechanical treatment, 240
  twin screw extruder, 104, 240

Fibre, 128, 147, 166, 199, 203, 215

analysis, 124, 151
barley, 203
fecal transit time, 206
fortification, 209
functional properties, 203
hypocholesterolemic effect, 209
insoluble, 124, 209, 217
oat, 203
physiological effects, 123, 150,
  203
soluble, 124, 150
Food, 43, 76, 121, 127, 137, 147,
              157, 191, 203, 240, 252,
              280
  food form, 130, 137
  functional food, 177
    caloric reduction, 164
    digesting stimulation, 177
    disease prevention, 206
  starchy food, 122, 127, 137
Food labelling, 121

Galactomannan, 76, 148, 191
Gelation, 86, 159, 230, 278
Gels, 4, 19, 76, 124, 159, 192, 221,
              278
  crystallinity, 278
  melting enthalpy, 286
  shear modulus, 278
  thermal reversibility, 221, 278
  two-phasic systems, 278
Glass transition, 63
  temperature, 175, 263
β-Glucanase, 26, 204
  (1→3)-β-glucanase, 26
  (1→4)-β-glucanase, 26
β-Glucans, 26, 132, 148, 180, 204,
              220
  (1→3)-β-glucan, 26, 220
  (1→4)-β-glucan, 26, 220
Glucanohydrolase, 26
Glucoamylase, 16, 234
Glucose, 6, 17, 36, 121, 127, 137,
              149, 192, 205, 253
  rapidly available glucose, 139
Glutathionylation, 237
Glycaemic index, 127, 139, 150
Glycaemic response, 122, 127, 149
Glycosidic linkage, 3, 215
Graphic paper, 180
  abrasion resistance, 187
  coatings, 180

gloss, 171, 187
printability, 186
smoothness, 187
Guar dispersions, 58
Gums, 53, 121, 195
cassia, 191
guar, 43, 76, 149, 191
locust bean, 76, 191
tara, 76, 191
xanthan, 76, 201

Hemicellulase, 104
Hemicellulose, 5, 199, 205, 215, 253
Hydrocyclone techniques, 180, 204
counter current washing, 181
fractionation, 181
purification, 181
Hydrolysis, 3, 21, 26, 133, 137, 148,
192, 233, 241, 278
index, 134
Hydroxyethylation, 262
Hypocholesterolemia, 209

Ileostomy, 143
Inhibition of gene expression, 36
Insulin, 122, 127, 149
Inulin, 121
Isoamylase, 16
Isoenzyme E I, 26
Isoenzyme E II, 26

Kappa carrageenan, 199

Legumes, 127, 138, 152
Leguminosae, 191
Lignin, 123, 147, 168, 199, 205, 252
Lignocellulose, 252
Limit dextrin, 16
Long range conformation, 73
Low-density lipoprotein
cholesterol, 208

Maltodextrin, 163, 278
molecular parameters, 280
sol-gel transition, 278
solution, 279
Mechanical spectrometer, 63, 104
Metabolic disease, 147
colorectal cancer, 147
diverticulitis, 147, 207
haemorrhoids, 147
Metabolism, 31, 123, 144, 149

carbohydrates, 149
cholesterol, 149
Monosaccharides, 121, 192
Muffins, 210

NMR, 3, 279
free induction decay, 279
low resolution, 284
solid-liquid ratio, 280
Non-starch polysaccharides, 121,
138, 147
Nondestructive measurements, 278
Nutritional classification of starch,
121, 127, 138
rapidly digestible starch, 138
resistant starch, 124, 132, 137,
148, 163
physically inaccessible, 138
resistant starch granules, 137
retrograded starch, 128, 138
slowly digestible starch, 138

Oats, 131, 142, 149, 180, 203, 220
fractionation of, 184
oat bran, 131, 150, 207
Oligosaccharides, 4, 121, 233, 280
Oscillatory rheological properties, 43

Parallel-plate, 117
Particle size distribution, 180, 253
laser diffraction spectrometry,
182
Pasta, 127, 137
Pectinase, 4
Pectins, 3, 44, 76, 121, 148, 215
Penetrant, 262
Plasticizer, 63, 265
Polymeric dispersions, 62
Polymers, 262
rubber, 265
swelling of, 262
Polysaccharides, 3, 26, 55, 76, 121,
137, 147, 158, 191, 215,
237, 278
chain conformation, 215
fringed micelle, 215
hydrodynamic radius, 222
lateral (side-by-side) aggregation,
215
radius of gyration, 215
specific linkage, 216

static and dynamic light
        scattering, 215
    suprastructure, 215
Porridge, 130, 137
Power-law, 68, 262
Protease, 104, 152, 237
Protein-carbohydrate
        interactions, 234,

Random coil molecules, 44
Regression analysis, 60, 98
Relaxation time, 44, 79, 279
Rhamnogalacturonan, 5
Rheological constants, 59
Rheology, 43, 78, 87, 159, 215, 285
    of wheat flour doughs, 43, 104
        elastomeric ..., 104
        viscoelastic ..., 43
Rheometry, 68, 104
Rhizopus arrhizus, 19, 31
Rigid rod model, 45

Serotonin, 123
Serum lipids, 203
Shear rate, 48, 105, 184, 241
Short range flexibility, 74
Slip-link network model, 50
Slit die, 69, 104
    slip, 104
    slippage, 104
Slit rheometry, 68
Span length, 19
Starch based binders, 166
    carbon skeleton, 167
    carbonisation, 167
    green bonding, 167
    hot bonding, 167
Starch containing emulsions, 171
Starch containing polymers, 168
    emulsions, 168
    solid polymers, 168
    soluble polymers, 168
Starch granules, 16, 131, 137, 158,
        180, 205, 234, 241, 264
    agglomeration of, 185
    granule swelling, 161, 184
    small granule starch (SGS), 180
    small starch granules (SSG), 180
    water surface interaction, 158,
        180
Starch molecules, 159, 180

amylopectin, 16, 37, 121, 128,
        158, 231, 233, 278
amylose, 16, 38, 121, 128, 137,
        148, 158, 215, 246, 278
    retrograded, 128, 138, 148,
        165, 286
cycloheptaamylose, 235
interaction, 128, 158, 180, 216,
        278
self aggregation, 289
Starch synthase, 36
Starch, 16, 33, 87, 121, 127, 137,
        148, 157, 180, 192, 203,
        233, 240, 262, 278
    amaranth, 180
    amylose free, 38
    barley, 131, 182, 203
    cationization, 245
    chemical substitution, 160
    consumption, 158
    crosslinked starches, 171
        for UHT processes, 160
    crystal structure, 137
    crystallinity, 128, 137, 278
    derivates, 162
    dicarboxylic, 169
    digestibility, 123, 138
    emulsifying starches, 162
    fat replacers, 163
        thermoreversible gels, 164
    fibrids (Chart-Bi), 173
    gelatinisation/gelation, 159
    maize, 87, 142, 161, 184
    modification, 159, 240
    native, 95, 160, 180, 203, 242
    n-octenyl succinate starches, 162
    oat, 131, 180, 203
    pastes, 158, 243
    plastification, 175
    poly-carboxylic-copolymers, 169
    potato, 127, 137, 163, 184, 203,
        262, 280
    production, 157
    super absorbent polymers, 171
    thermostability, 240
    water binding capacity, 185
    wheat, 92, 131, 141, 180, 192,
        203
Starch/latex mixtures, 170
Starch-less potatoes, 37
Steady viscosity, 64
Storage modulus, 44, 78

Thickening agent, 191
    drilling fluids, 191
    oil recovery, 191
    paper coatings, 168, 186
    printing ink, 191
Tissue-specific expression, 29
Transgenic plant, 33

Viscosity, 44, 76, 87, 104, 149, 158,
            180, 191, 204, 222, 241,
            278
    Brookfield, 184
    intrinsic, 52, 77, 180, 222
    shear rate measurement, 63

Water, 87, 105, 185, 205, 206, 271
    binding capacity, 185, 206
    content, 87, 105, 271
    holding capacity, 206
    up-take, 205

X-ray crystallography, 282
Xanthan/galactomannan mixtures, 77
Xylanase, 4
Xylans, 144, 148, 226
Xylogalacturonan, 7